U0344319

中国
城乡建设统计年鉴

中华人民共和国住房和城乡建设部 编

Ministry of Housing and Urban-Rural Development, P.R.CHINA

2022

China
Urban-Rural
Construction
Statistical
Yearbook

中国统计出版社

China Statistics Press

图书在版编目（CIP）数据

中国城乡建设统计年鉴. 2022 = China Urban-Rural Construction Statistical Yearbook 2022：汉英对照 / 中华人民共和国住房和城乡建设部编. -- 北京 ：中国统计出版社, 2023.9

ISBN 978-7-5230-0171-4

Ⅰ. ①中… Ⅱ. ①中… Ⅲ. ①城乡建设－统计资料－中国－2022－年鉴－汉、英 Ⅳ. ①TU984.2-54

中国国家版本馆 CIP 数据核字(2023)第 141083 号

中国城乡建设统计年鉴—2022

作　　者/中华人民共和国住房和城乡建设部编
责任编辑/钟　钰
封面设计/李　静
出版发行/中国统计出版社有限公司
通信地址/北京市丰台区西三环南路甲 6 号　邮政编码/100073
发行电话/邮购（010）63376909　书店（010）68783171
网　　址/http://www.zgtjcbs.com/
印　　刷/河北鑫兆源印刷有限公司
经　　销/新华书店
开　　本/890mm×1240mm　1/16
字　　数/432 千字
印　　张/14.5
版　　别/2023 年 9 月第 1 版
版　　次/2023 年 9 月第 1 次印刷
定　　价/168.00 元

版权所有，侵权必究。

《中国城乡建设统计年鉴—2022》

编委会和编辑工作人员

一、编委会

主　　任：姜万荣

副主任：宋友春

编　　委：（以地区排名为序）

张铁军	杨　琦	侯学钢	王　伟	张　强	徐向东	符里刚	宋　刚
揭新民	聂海俊	曹桂喆	王光辉	邢文忠	刘晓东	徐东锋	金　晨
马　韧	陈浩东	路宏伟	张清云	姚昭晖	刘孝华	宋直刚	朱子君
苏友佺	李道鹏	熊春华	王玉志	董海立	金　涛	谈华初	吴启旺
易小林	李海平	杨绿峰	汪夏明	陈积卫	陈光宇	张其悦	叶长春
陈福均	陈　勇	边　疆	赵志勇	李修武	李进忠	付　涛	王　勇
李兰宏	张志杰	马成贵	李　斌	李林毓	木塔力甫·艾力		王恩茂

二、编辑工作人员

总　编　辑：南　昌

副总编辑：卢　嘉　李　威　李雪娇

编辑人员：（以地区排名为序）

郑　炎	王宇慧	诸葛绪光	石　林	张志嵩	张利兵	朱芃莉	付兴华
张　伟	李芳芳	温炎涛	姚　娜	杜艳捷	王玉琦	刘　勇	杨　婧
王小飞	李颖慧	梁立方	曹亚群	冷会杰	司　慧	沈晓红	付肇群
宫新博	孙辉东	王　欢	林　岩	肖楚宇	王嘉琦	王　爽	孙莉利
谷洪义	杜金芝	苑晓东	邓绪明	江　星	樊　好	路文龙	王　青
张　力	丁　化	陈明伟	俞　非	王山红	贾利松	苏　娟	吴毅峰
邱亦东	胡　璞	沈昆卉	卢立新	孟　奎	余　燕	陈文杰	王少彬
王晓霞	黄新华	王登平	陈善游	姚　坡	辛　颖	王建鑫	常旭阳
刘二鹏	董雨辰	郭彩文	李洪涛	张　煦	韩文超	张　雷	古艳艳
查良春	禹滋柏	凌小刚	周　旭	田明革	叶　昂	徐碧波	尹清源
杨爱春	赵鹏凯	朱文静	吴　茵	曾俊杰	黄玉清	高　磊	秦德坤
罗　翼	叶　茂	林远征	廖　楠	江　浩	叶晓璇	裴　玮	刘宇飞
王建欣	林　琳	安旭慧	莫志刚	梅朝伟	彭　里	沈　键	李平辉
马　望	姚文锋	格桑顿珠	熊艳玲	平措桑珠	德庆卓嘎		张立群
杨　莹	王光辉	毕东涛	马筵栋	张智博	于学刚	王章军	李　崑
李　军	马路遥	吕建华	马　玉	杨　帆	巴尔古丽·依明		张　辉
李璠璠							

China Urban-Rural Construction Statistical Yearbook—2022
Editorial Board and Editorial Staff

编辑说明

一、为全面反映我国城乡市政公用设施建设与发展状况，方便国内外各界了解中国城乡建设全貌，我们编辑了《中国城乡建设统计年鉴》《中国城市建设统计年鉴》和《中国县城建设统计年鉴》中英文对照本，每年公布一次，供社会广大读者作为资料性书籍使用。

二、本年鉴的统计范围

设市城市的城区：市本级（1）街道办事处所辖地域；（2）城市公共设施、居住设施和市政公用设施等连接到的其他镇（乡）地域；（3）常住人口在3000人以上独立的工矿区、开发区、科研单位、大专院校等特殊区域。

县城：（1）县政府驻地的镇、乡或街道办事处地域（城关镇）；（2）县城公共设施、居住设施和市政公用设施等连接到的其他镇（乡）地域；（3）常住人口在3000人以上独立的工矿区、开发区、科研单位、大专院校等特殊区域。

村镇：政府驻地的公共设施和居住设施没有和城区（县城）连接的建制镇、乡和镇乡级特殊区域。

三、2022年底全国境内31个省、自治区、直辖市，共有691个设市城市，1472个县（含自治县、旗、自治旗、林区、特区。下同），21387个建制镇，8229个乡（含民族乡、苏木、民族苏木。下同），48.9万个行政村。

四、本年鉴根据各省、自治区和直辖市建设行政主管部门上报的2022年城乡建设统计数据编辑，由城市（城区）、县城、村镇三部分组成：

（一）城市（城区）部分，统计了690个城市、5个特殊区域。5个特殊区域包括辽宁省沈抚改革创新示范区、吉林省长白山保护开发区管理委员会、河南省郑州航空港经济综合实验区、陕西省杨凌区和宁夏回族自治区宁东。

（二）县城部分，统计了1466个县和15个特殊区域。河北省邢台县、沧县，山西省泽州县，辽宁省抚顺县、铁岭县，新疆维吾尔自治区乌鲁木齐县、和田县等7个县，因与所在城市市县同城，县城部分不含上述县城数据，数据含在其所在城市中；福建省金门县暂无数据资料。

15个特殊区域包括河北省曹妃甸区、白沟新城，山西省云州区，黑龙江省加格达奇区，湖北省神农架林区，湖南省望城区、南岳区、大通湖区，海南省洋浦经济开发区，四川省东部新区管理委员会，贵州省六枝特区，云南省昆明阳宗海风景名胜区，青海省西海镇、大柴旦行委，宁夏回族自治区红寺堡区。

（三）村镇部分，统计了 19245 个建制镇、7959 个乡、407 个镇乡级特殊区域和 233.2 万个自然村（其中村民委员会所在地 47.8 万个）。

五、本年鉴数据不包括香港特别行政区、澳门特别行政区以及台湾省。

六、本年鉴中除人均住宅建筑面积、人均日生活用水外，所有人均指标、普及率指标均以户籍人口与暂住人口合计为分母计算。

七、本年鉴中"空格"表示该项统计指标数据不足本表最小单位数、数据不详或无该项数据。

八、本年鉴中部分数据合计数或相对数由于单位取舍不同而产生的计算误差，均没有进行机械调整。

九、为促进中国建设行业统计信息工作的发展，欢迎广大读者提出改进意见。

EDITOR'S NOTES

1. *China Urban-Rural Construction Statistical Yearbook*, *China Urban Construction Statistical Yearbook* and *China County Seat Construction Statistical Yearbook* are published annually in both Chinese and English languages to provide comprehensive information on urban and rural service facilities development in China. Being the source of facts, the yearbooks help to facilitate the understanding of people from all walks of life at home and abroad on China's urban and rural development.

2. Coverage of the statistics

Urban Areas: (1) areas under the jurisdiction of neighborhood administration; (2) other towns (townships) connected to urban public facilities, residential facilities and municipal utilities; (3) special areas like independent industrial and mining districts, development zones, research institutes, and universities and colleges with permanent residents of 3000 and above.

County Seat Areas: (1) towns and townships where county governments are situated and areas under the jurisdiction of neighborhood administration; (2) other towns (townships) connected to county seat public facilities, residential facilities and municipal utilities; (3) special areas like independent industrial and mining districts, development zones, research institutes, and universities and colleges with permanent residents of 3000 and above.

Villages and Small Towns Areas: towns, townships and special district at township level of which public facilities and residential facilities are not connected to those of cities (county seats).

3. There were a total of 691 cities, 1472 counties (including autonomous counties, banners, autonomous banners, forest districts, and special districts), 21387 towns, 8229 townships (including minority townships, Sumus and minority Sumus), and 489 thousand administrative villages in all the 31 provinces, autonomous regions and municipalities across China by the end of 2022.

4. The yearbook is compiled based on statistical data on urban and rural construction in year 2022 that were reported by construction authorities of provinces, autonomous regions and municipalities directly under the central government. The yearbook is composed of statistics for three parts, namely statistics for cities (urban districts), county seats, and villages and small towns.

(1) In the part of Cities (Urban Districts), data are from 690 cities, 5 special zones. Shenfu Reform and Innovation Demonstration Zone in Liaoning Province, Changbai Mountain Protection Development Management Committee in Jilin Province, Zhengzhou Airport Economy Zone in Henan Province, Yangling District in Shanxi Province and Ningdong in ningxia Autonomous Region are classified as city.

(2) In the part of County Seats, data is from 1466 counties, 15 special zones and districts. Data from 7 counties including Xingtai and Cangxian County in Hebei Province, Zezhou County in Shanxi Province, Fushun and Tieling County in Liaoning Province, and Urumqi and Hetian County in Xinjiang Uygur Autonomous Region are included in the statistics of the respective cities administering the above counties due to the identity of the location between the county seats and the cities; data on Jinmen County in Fujian Province has not been available at the moment.

15 special zones and districts include Caofeidian and Baigouxincheng in Hebei Province, Yunzhou District in

Shanxi Province, Jiagedaqi in Heilongjiang Province, Shennongjia Forestry District in Hubei Province, Wangcheng District, Nanyue District and Datong Lake District in Hunan Province, Yangpu Economic Development Zone in Hainan Province, Eastern New Area in Sichuan Province, Liuzhite District in Guizhou Province, Kunming Yangzonghai Scenic Spot in Yunnan Province, Xihai Town and Dachaidan Administrative Commission in Qinghai Province, and Hongsibao Development Zone in Ningxia Autonomous Region.

(3) In the part of Villages and Towns, statistics is based on data from 19245 towns, 7959 townships, 407 special areas at the level of town and township, and 2.332 million natural villages including 478 thousand villages where villagers' committees are situated.

5. This Yearbook does not include data of Hong Kong Special Administrative Region, Macao Special Administrative Region and Taiwan Province.

6. All the per capita and coverage rate data in this Yearbook, except the per capita residential floor area and per capita daily consumption of domestic water, are calculated using the sum of resident and non-resident population as a denominator.

7. In this yearbook "blank space" indicates that the figure is not large enough to be measured with the smallest unit in the table,or data are unknown or are not available.

8. The calculation errors of the total or relative value of some data in this Yearbook arising from the use of different measurement units have not been mechanically aligned.

9. Any comments to improve the quality of the yearbook are welcomed to promote the advancement in statistics in China's construction industry.

目 录
Contents

城市部分
Statistics for Cities

一、综合数据
General Data

1-1-1　全国历年城市市政公用设施水平 ·· 3
General Data Level of National Urban Service Facilities in Past Years

1-1-2　全国城市市政公用设施水平(2022 年) ·· 4
National Level of National Urban Service Facilities(2022)

1-2-1　全国历年城市数量及人口、面积情况 ·· 6
National Changes in Number of Cities, Urban Population and Urban Area in Past Years

1-2-2　全国城市人口和建设用地(2022 年) ·· 8
National Urban Population and Construction Land (2022)

1-3-1　全国历年城市维护建设资金收入 ·· 10
National Revenue of Urban Maintenance and Construction Fund in Past Years

1-3-2　全国历年城市维护建设资金支出 ·· 12
National Expenditure of Urban Maintenance and Construction Fund in Past Years

1-4-1　按行业分全国历年城市市政公用设施建设固定资产投资 ············ 14
National Fixed Assets Investment in Urban Service Facilities by Industry in Past Years

1-4-2　按行业分全国城市市政公用设施建设固定资产投资(2022 年) ····· 16
National Fixed Assets Investment in Urban Service Facilities by Industry(2022)

1-5-1　按资金来源分全国历年城市市政公用设施建设固定资产投资 ········· 18
National Fixed Assets Investment in Urban Service Facilities by Capital Source in Past Years

1-5-2　按资金来源分全国城市市政公用设施建设固定资产投资(2022 年) ··· 20
National Fixed Assets Investment in Urban Service Facilities by Capital Source(2022)

二、居民生活数据
Data by Residents Living

1-6-1　全国历年城市供水情况 ·· 22
National Urban Water Supply in Past Years

1-6-2　城市供水(2022 年) ··· 24
Urban Water Supply(2022)

1-6-3　城市供水(公共供水)(2022 年) ··· 26
Urban Water Supply(Public Water Suppliers)(2022)

1-6-4　城市供水(自建设施供水)(2022 年) ···················· 28
　　　　Urban Water Supply (Suppliers with Self-Built Facilities)(2022)

1-7-1　全国历年城市节约用水情况 ···························· 30
　　　　National Urban Water Conservation in Past Years

1-7-2　城市节约用水 (2022 年) ······························ 32
　　　　Urban Water Conservation (2022)

1-8-1　全国历年城市燃气情况 ································ 34
　　　　National Urban Gas in Past Years

1-8-2　城市人工煤气(2022 年) ······························ 36
　　　　Urban Man-Made Coal Gas(2022)

1-8-3　城市天然气 (2022 年) ·································· 38
　　　　Urban Natural Gas(2022)

1-8-4　城市液化石油气(2022 年) ···························· 40
　　　　Urban LPG Supply(2022)

1-9-1　全国历年城市集中供热情况 ···························· 42
　　　　National Urban Centralized Heating in Past Years

1-9-2　城市集中供热(2022 年) ······························ 44
　　　　Urban Central Heating(2022)

三、居民出行数据
Data by Residents Travel

1-10-1　全国历年城市轨道交通情况 ·························· 46
　　　　Data by Residents Travel National Urban Rail Transit System in Past Years

1-10-2　城市轨道交通(建成)(2022 年) ······················ 48
　　　　Urban Rail Transit System (Completed)(2022)

1-10-3　城市轨道交通(在建)(2022 年) ······················ 50
　　　　Urban Rail Transit System(Under Construction) (2022)

1-11-1　全国历年城市道路和桥梁情况 ························ 52
　　　　National Urban Road and Bridge in Past Years

1-11-2　城市道路和桥梁 (2022 年) ·························· 54
　　　　Urban Roads and Bridges(2022)

四、环境卫生数据
Data by Environmental Health

1-12-1　全国历年城市排水和污水处理情况 ···················· 56
　　　　Data by Environmental Health of Residents National Urban Drainage and Wastewater Treatment in Past Years

1-12-2　城市排水和污水处理(2022 年) ······················ 58
　　　　Urban Drainage and Wastewater Treatment(2022)

1-13-1　全国历年城市市容环境卫生情况 ······················ 60
　　　　National Urban Environmental Sanitation in Past Years

1-13-2　城市市容环境卫生(2022 年) ························ 62
　　　　Urban Environmental Sanitation(2022)

五、绿色生态数据
Data by Green Ecology

    1-14-1   全国历年城市园林绿化情况 ·· 64

                Data by Green Ecological Environment for Residents National Urban Landscaping

                in Past Years

    1-14-2   城市园林绿化(2022 年) ·· 66

                Urban Landscaping(2022)

县城部分
Statistics for County Seats

一、综合数据
General Data

    2-1-1   全国历年县城市政公用设施水平 ·· 71

                General Data Level of Service Facilities of National County Seat in Past Years

    2-1-2   全国县城市政公用设施水平(2022 年) ·· 72

                Level of National County Seat Service Facilities(2022)

    2-2-1   全国历年县城数量及人口、面积情况 ··· 74

                National Changes in Number of Counties, Population and Area in Past Years

    2-2-2   全国县城人口和建设用地(2022 年) ··· 76

                County Seat Population and Construction Land(2022)

    2-3-1   全国历年县城维护建设资金收入 ·· 78

                National Revenue of County Seat Maintenance and Construction Fund in Past Years

    2-3-2   全国历年县城维护建设资金支出 ·· 80

                National Expenditure of County Seat Maintenance and Construction Fund in Past Years

    2-4-1   按行业分全国历年县城市政公用设施建设固定资产投资 ·· 82

                National Fixed Assets Investment of County Seat Service Facilities by Industry in Past Years

    2-4-2   按行业分全国县城市政公用设施建设固定资产投资(2022 年) ···································· 84

                National Investment in Fixed Assets of County Seat Service Facilities by Industry(2022)

    2-5-1   按资金来源分全国历年县城市政公用设施建设固定资产投资 ······································ 86

                National Fixed Assets Investment of County Seat Service Facilities by Capital Source

                in Past Years

    2-5-2   按资金来源分全国县城市政公用设施建设固定资产投资(2022 年) ······························· 88

                National Investment in Fixed Assets of County Seat Service Facilities by Capital Source (2022)

二、居民生活数据
Data by Residents Living

    2-6-1   全国历年县城供水情况 ·· 90

                National County Seat Water Supply in Past Years

    2-6-2   县城供水(2022 年) ·· 92

                County Seat Water Supply(2022)

    2-6-3   县城供水(公共供水)(2022 年) ··· 94

                County Seat Water Supply (Public Water Suppliers)(2022)

2-6-4 县城供水(自建设施供水)(2022 年) ·· 96
County Seat Water Supply (Suppliers with Self-Built Facilities)(2022)

2-7-1 全国历年县城节约用水情况 ··· 98
National County Seat Water Conservation in Past Years

2-7-2 县城节约用水(2022 年) ·· 100
County Seat Water Conservation(2022)

2-8-1 全国历年县城燃气情况 ··· 102
National County Seat Gas in Past Years

2-8-2 县城人工煤气(2022 年) ·· 104
County Seat Man-Made Coal Gas(2022)

2-8-3 县城天然气(2022 年) ··· 106
County Seat Natural Gas(2022)

2-8-4 县城液化石油气(2022 年) ··· 108
County Seat LPG Supply (2022)

2-9-1 全国历年县城集中供热情况 ··· 110
National County Seat Centralized Heating in Past Years

2-9-2 县城集中供热(2022 年) ·· 112
County Seat Central Heating(2022)

三、居民出行数据
Data by Residents Travel

2-10-1 全国历年县城道路和桥梁情况 ··· 114
Data by Residents Travel National County Seat Road and Bridge in Past Years

2-10-2 县城道路和桥梁(2022 年) ··· 115
County Seat Roads and Bridges(2022)

四、环境卫生数据
Data by Environmental Health

2-11-1 全国历年县城排水和污水处理情况 ··· 117
Data by Environmental Health of Residents National County Seat Drainage and
Wastewater Treatment in Past Years

2-11-2 县城排水和污水处理(2022 年) ·· 118
County Seat Drainage and Wastewater Treatment(2022)

2-12-1 全国历年县城市容环境卫生情况 ·· 120
National County Seat Environmental Sanitation in Past Years

2-12-2 县城市容环境卫生(2022 年) ··· 122
County Seat Environmental Sanitation(2022)

五、绿色生态数据
Data by Green Ecology

2-13-1 全国历年县城园林绿化情况 ··· 124
National County Seat Landscaping in Past Years

2-13-2 县城园林绿化(2022 年) ··· 126
County Seat Landscaping(2022)

村镇部分

Statistics for Villages and Small Towns

3-1-1　全国历年建制镇及住宅基本情况 ·································· 130
National Summary of Towns and Residential Building in Past Years

3-1-2　全国历年建制镇市政公用设施情况 ···························· 132
National Municipal Public Facilities of Towns in Past Years

3-1-3　全国历年乡及住宅基本情况 ···································· 134
National Summary of Townships and Residential Building in Past Years

3-1-4　全国历年乡市政公用设施情况 ·································· 136
National Municipal Public Facilities of Townships in Past Years

3-1-5　全国历年村庄基本情况 ·· 138
National Summary of Villages in Past Years

3-2-1　建制镇市政公用设施水平(2022 年) ····························· 140
Level of Municipal Public Facilities of Built-up Area of Towns (2022)

3-2-2　建制镇基本情况(2022 年) ····································· 142
Summary of Towns(2022)

3-2-3　建制镇供水(2022 年) ·· 144
Water Supply of Towns(2022)

3-2-4　建制镇燃气、供热、道路桥梁(2022 年) ························ 146
Gas, Central Heating, Road and Bridges of Towns(2022)

3-2-5　建制镇排水和污水处理(2022 年) ······························ 148
Drainage and Wastewater Treatment of Towns(2022)

3-2-6　建制镇园林绿化及环境卫生(2022 年) ·························· 149
Landscaping and Environmental Sanitation of Towns(2022)

3-2-7　建制镇房屋(2022 年) ·· 150
Building Construction of Towns(2022)

3-2-8　建制镇建设投入(2022 年) ····································· 152
Construction Input of Towns(2022)

3-2-9　乡市政公用设施水平(2022 年) ································· 154
Level of Municipal Public Facilities of Built-up Area of Townships(2022)

3-2-10　乡基本情况(2022 年) ·· 156
Summary of Townships(2022)

3-2-11　乡供水(2022 年) ·· 158
Water Supply of Townships(2022)

3-2-12　乡燃气、供热、道路桥梁(2022 年) ··························· 160
Gas, Central Heating, Roads and Bridges of Townships(2022)

3-2-13　乡排水和污水处理(2022 年) ·································· 162
Drainage and Wastewater Treatment of Townships(2022)

3-2-14　乡园林绿化及环境卫生(2022 年) ····························· 163
Landscaping and Environmental Sanitation of Townships(2022)

3-2-15　乡房屋(2022 年)···································164
　　　　Building Construction of Townships(2022)

3-2-16　乡建设投入(2022 年)·······························166
　　　　Construction Input of Townships(2022)

3-2-17　镇乡级特殊区域市政公用设施水平(2022 年)···············168
　　　　Level of Municipal Public Facilities of Built-up Area of Special District at Township Level
　　　　(2022)

3-2-18　镇乡级特殊区域基本情况(2022 年)·····················170
　　　　Summary of Special District at Township Level(2022)

3-2-19　镇乡级特殊区域供水(2022 年)·························172
　　　　Water Supply of Special District at Township Level(2022)

3-2-20　镇乡级特殊区域燃气、供热、道路桥梁(2022 年)·············174
　　　　Gas, Central Heating, Road and Bridge of Special District at Township Level(2022)

3-2-21　镇乡级特殊区域排水和污水处理(2022 年)·················176
　　　　Drainage and Wastewater Treatment of Special District at Township Level(2022)

3-2-22　镇乡级特殊区域园林绿化及环境卫生(2022 年)···············177
　　　　Landscaping and Environmental Sanitation of Special District at Township Level(2022)

3-2-23　镇乡级特殊区域房屋(2022 年)·························178
　　　　Building Construction of Special District at Township Level(2022)

3-2-24　镇乡级特殊区域建设投入(2022 年)·····················180
　　　　Construction Input of Special District at Township Level(2022)

3-2-25　村庄人口及面积(2022 年)···························182
　　　　Population and Area of Villages(2022)

3-2-26　村庄公共设施(一)(2022 年)·························184
　　　　Public Facilities of Villages Ⅰ(2022)

3-2-27　村庄公共设施(二)(2022 年)·························186
　　　　Public Facilities of Villages Ⅱ(2022)

3-2-28　村庄房屋(2022 年)·······························188
　　　　Building Construction of Villages(2022)

3-2-29　村庄建设投入(2022 年)·····························190
　　　　Construction Input of Villages(2022)

主要指标解释···195
Explanatory Notes on Main Indicators

城市部分

Statistics for Cities

一、综合数据
General Data

1-1-1　全国历年城市市政公用设施水平
Level of National Urban Service Facilities in Past Years

年份 Year	供水普及率 (%) Water Coverage Rate (%)	燃气普及率 (%) Gas Coverage Rate (%)	每万人拥有公共交通车辆(标台) Motor Vehicle for Public Transport Per 10,000 Persons (standard unit)	人均道路面积(平方米) Road Surface Area Per Capita (m²)	污水处理率 (%) Wastewater Treatment Rate (%)	园林绿化 Landscaping			每万人拥有公厕(座) Number of Public Lavatories per 10,000 Persons (unit)
						人均公园绿地面积(平方米) Public Recreational Green Space Per Capita (m²)	建成区绿化覆盖率 (%) Green Coverage Rate of Built District (%)	建成区绿地率 (%) Green Space Rate of Built District (%)	
1978									
1979									
1980									
1981	53.7	11.6		1.81		1.50			3.77
1982	56.7	12.6		1.96		1.65			3.99
1983	52.5	12.3		1.88		1.71			3.95
1984	49.5	13.0		1.84		1.62			3.57
1985	45.1	13.0		1.72		1.57			3.28
1986	51.3	15.2	2.5	3.05		1.84	16.90		3.61
1987	50.4	16.7	2.4	3.10		1.90	17.10		3.54
1988	47.6	16.5	2.2	3.10		1.76	17.00		3.14
1989	47.4	17.8	2.1	3.22		1.69	17.80		3.09
1990	48.0	19.1	2.2	3.13		1.78	19.20		2.97
1991	54.8	23.7	2.7	3.35	14.86	2.07	20.10		3.38
1992	56.2	26.3	3.0	3.59	17.29	2.13	21.00		3.09
1993	55.2	27.9	3.0	3.70	20.02	2.16	21.30		2.89
1994	56.0	30.4	3.0	3.84	17.10	2.29	22.10		2.69
1995	58.7	34.3	3.6	4.36	19.69	2.49	23.90		3.00
1996	60.7	38.2	3.8	4.96	23.62	2.76	24.43	19.05	3.02
1997	61.2	40.0	4.5	5.22	25.84	2.93	25.53	20.57	2.95
1998	61.9	41.8	4.6	5.51	29.56	3.22	26.56	21.81	2.89
1999	63.5	43.8	5.0	5.91	31.93	3.51	27.58	23.03	2.85
2000	63.9	45.4	5.3	6.13	34.25	3.69	28.15	23.67	2.74
2001	72.26	60.42	6.10	6.98	36.43	4.56	28.38	24.26	3.01
2002	77.85	67.17	6.73	7.87	39.97	5.36	29.75	25.80	3.15
2003	86.15	76.74	7.66	9.34	42.39	6.49	31.15	27.26	3.18
2004	88.85	81.53	8.41	10.34	45.67	7.39	31.66	27.72	3.21
2005	91.09	82.08	8.62	10.92	51.95	7.89	32.54	28.51	3.20
2006	86.07 (97.04)	79.11 (88.58)	9.05 (10.13)	11.04 (12.36)	55.67	8.30 (9.30)	35.11	30.92	2.88 (3.22)
2007	93.83	87.40	10.23	11.43	62.87	8.98	35.29	31.30	3.04
2008	94.73	89.55	11.13	12.21	70.16	9.71	37.37	33.29	3.12
2009	96.12	91.41		12.79	75.25	10.66	38.22	34.17	3.15
2010	96.68	92.04		13.21	82.31	11.18	38.62	34.47	3.02
2011	97.04	92.41		13.75	83.63	11.80	39.22	35.27	2.95
2012	97.16	93.15		14.39	87.30	12.26	39.59	35.72	2.89
2013	97.56	94.25		14.87	89.34	12.64	39.70	35.78	2.83
2014	97.64	94.57		15.34	90.18	13.08	40.22	36.29	2.79
2015	98.07	95.30		15.60	91.90	13.35	40.12	36.36	2.75
2016	98.42	95.75		15.80	93.44	13.70	40.30	36.43	2.72
2017	98.30	96.26		16.05	94.54	14.01	40.91	37.11	2.77
2018	98.36	96.70		16.70	95.49	14.11	41.11	37.34	2.88
2019	98.78	97.29		17.36	96.81	14.36	41.51	37.63	2.93
2020	98.99	97.87		18.04	97.53	14.78	42.06	38.24	3.07
2021	99.38	98.04		18.84	97.89	14.87	42.42	38.70	3.29
2022	99.39	98.06		19.28	98.11	15.29	42.96	39.29	3.43

注：1. 自2006年起，人均和普及率指标按城区人口和城区暂住人口合计为分母计算，以公安部门的户籍统计和暂住人口统计为准。括号中的数据为与往年同口径数据。

2. "人均公园绿地面积"指标2005年及以前年份为"人均公共绿地面积"。

3. 自2009年起，城市公共交通内容不再统计，增加城市轨道交通建设情况内容。

Notes: 1. Since 2006, figure in terms of per capita and coverage rate have been calculated based on denominater which combines both permanent and temporary residents in urban areas. And the population should come from statistics of police. The data in brackets are same index calculated by the method of past years.

2. Since 2005, Public Green Space Per Capita is changed to Public Recreational Green Space Per Capita.

3. Since 2009, statistics on urban public transport have been removed, and relevant information on the construction of rail transit system has been added.

1-1-2　全国城市市政公用设施水平(2022年)

地区名称 Name of Regions	人口密度 （人/平方公里） Population Density (person/square kilometer)	人均日生活用水量（升） Daily Water Consumption Per Capita (liter)	供水普及率（%） Water Coverage Rate (%)	公共供水普及率 Public Water Coverage Rate	燃气普及率（%） Gas Coverage Rate (%)	建成区供水管道密度（公里/平方公里） Density of Water Supply Pipelines in Built District (kilometer/square kilometer)	人均道路面积（平方米） Road Surface Area Per Capita (m²)	建成区路网密度（公里/平方公里） Road in Built District (kilometer/square kilometer)
全　国　National Total	2854	184.73	99.39	98.73	98.06	14.87	19.28	7.66
北　京　Beijing		163.22	99.81	96.15	100.00		8.04	
天　津　Tianjin	4372	122.81	100.00	100.00	100.00	16.92	16.11	6.58
河　北　Hebei	3150	123.31	100.00	99.61	99.55	10.03	21.20	8.39
山　西　Shanxi	3855	136.52	98.66	97.49	97.56	10.87	18.71	7.41
内蒙古　Inner Mongolia	2002	119.25	99.70	99.32	97.96	10.00	24.57	7.60
辽　宁　Liaoning	1792	155.81	98.99	97.69	97.73	13.51	19.37	7.80
吉　林　Jilin	2097	121.83	96.46	95.33	96.54	10.22	17.15	6.50
黑龙江　Heilongjiang	5361	127.28	99.10	98.28	93.22	13.56	16.63	7.48
上　海　Shanghai	3905	207.04	100.00	100.00	100.00	32.22	5.00	4.82
江　苏　Jiangsu	2156	211.54	100.00	99.95	99.92	18.93	25.66	9.33
浙　江　Zhejiang	2344	215.05	100.00	99.70	99.97	21.90	19.60	8.13
安　徽　Anhui	2744	194.31	99.76	99.10	99.42	13.94	24.57	7.91
福　建　Fujian	3492	228.88	99.97	99.96	99.67	17.62	21.96	8.21
江　西　Jiangxi	3647	222.48	99.37	99.18	98.82	16.78	25.92	8.05
山　东　Shandong	1724	126.22	99.92	99.01	99.47	10.16	26.45	8.20
河　南　Henan	4480	139.89	99.30	97.23	98.21	8.74	16.86	5.30
湖　北　Hubei	3056	199.93	99.93	99.47	99.50	18.98	19.67	8.22
湖　南　Hunan	4717	217.07	99.01	98.75	97.70	18.05	20.35	8.21
广　东　Guangdong	3856	241.53	99.74	99.70	98.61	18.33	15.02	7.60
广　西　Guangxi	2473	273.73	99.91	99.49	99.44	14.26	24.40	8.51
海　南　Hainan	2487	287.04	99.95	98.19	99.53	9.30	25.27	11.69
重　庆　Chongqing	2079	180.77	98.57	97.78	98.82	14.83	16.64	7.38
四　川　Sichuan	3670	195.31	97.18	97.11	96.55	16.42	18.28	8.07
贵　州　Guizhou	2092	175.80	98.88	98.87	93.26	18.31	26.69	8.15
云　南　Yunnan	3290	186.59	99.01	98.70	71.75	13.23	17.40	7.11
西　藏　Tibet	1516	245.66	99.70	99.70	74.26	10.90	22.03	4.49
陕　西　Shaanxi	5321	162.50	98.25	95.65	99.03	7.93	18.11	5.70
甘　肃　Gansu	3223	138.88	99.50	98.88	96.93	6.37	22.21	7.24
青　海　Qinghai	2893	174.90	99.57	98.39	94.72	12.99	19.58	5.81
宁　夏　Ningxia	3103	165.80	99.99	99.91	98.48	6.90	28.00	5.88
新　疆　Xinjiang	3915	158.89	99.70	99.41	98.97	8.20	22.51	6.04
新疆生产建设兵团　Xinjiang Production and Construction Corps	1456	207.81	97.88	97.88	95.96	8.60	34.37	6.93

注：本表中北京市的建成区绿化覆盖率和建成区绿地率均为该市调查面积内数据，全国城市建成区绿化覆盖率和建成区绿地率作适当修正。

National Level of National Urban Service Facilities(2022)

建成区道路面积率(%) Road Surface Area Rate of Built District (%)	建成区排水管道密度(公里/平方公里) Density of Sewers in Built District (kilometer/ square kilometer)	污水处理率(%) Wastewater Treatment Rate (%)	污水处理厂集中处理率 Centralized Treatment Rate of Wastewater Treatment Plants	城市生活污水集中收集率(%) Centralized collection rate of urban domestic sewage (%)	人均公园绿地面积(平方米) Public Recreational Green Space Per Capita (m²)	建成区绿化覆盖率(%) Green Coverage Rate of Built District (%)	建成区绿地率(%) Green Space Rate of Built District (%)	生活垃圾处理率(%) Domestic Garbage Treatment Rate (%)	生活垃圾无害化处理率 Domestic Garbage Harmless Treatment Rate	地区名称 Name of Regions
15.41	12.34	98.11	96.50	70.06	15.29	42.96	39.29	99.98	99.90	全 国
		98.07	96.18	88.68	16.63	49.77	47.05	100.00	100.00	北 京
12.39	18.56	98.45	97.35	82.44	9.98	38.41	35.49	100.00	100.00	天 津
17.67	9.78	99.08	99.08	81.72	15.35	43.78	40.01	100.00	100.00	河 北
17.57	10.55	98.48	97.58	71.07	13.72	44.02	40.14	100.00	100.00	山 西
17.25	11.15	97.56	97.56	76.45	19.47	41.94	38.96	99.99	99.99	内 蒙 古
14.74	7.37	98.03	97.28	63.18	13.39	40.88	38.63	99.58	99.58	辽 宁
11.92	7.67	97.84	97.84	72.98	14.45	42.68	38.92	100.00	100.00	吉 林
12.34	7.06	96.95	94.13	67.80	14.04	37.97	34.68	100.00	100.00	黑 龙 江
9.96	17.95	98.04	98.04	90.84	9.28	38.10	36.86	100.00	98.44	上 海
16.49	15.49	97.42	93.13	75.37	16.02	44.07	40.65	100.00	100.00	江 苏
15.97	14.78	98.14	96.76	74.44	13.79	42.13	38.24	100.00	100.00	浙 江
18.55	13.98	98.07	96.27	61.85	16.98	45.32	41.15	100.00	100.00	安 徽
16.60	11.47	98.57	93.46	60.87	15.28	44.07	40.68	100.00	100.00	福 建
16.91	12.67	97.61	96.91	49.93	17.01	46.63	43.28	100.00	100.00	江 西
16.77	12.00	98.53	98.39	72.27	18.18	43.75	39.72	100.00	100.00	山 东
12.63	8.96	99.53	99.53	77.12	15.60	40.30	35.52	99.99	99.68	河 南
16.54	13.27	97.57	92.13	55.62	15.38	42.92	39.42	100.00	100.00	湖 北
17.56	11.98	98.17	98.10	59.22	13.06	42.32	38.72	100.00	100.00	湖 南
13.48	16.40	98.54	98.06	72.41	17.95	44.56	40.25	100.00	99.95	广 东
16.03	12.28	98.78	92.93	54.54	11.67	42.21	36.86	100.00	100.00	广 西
18.58	13.53	99.15	98.89	55.94	12.23	42.39	39.43	100.00	100.00	海 南
15.99	15.16	98.36	98.06	63.25	17.63	44.56	41.33	100.00	100.00	重 庆
16.20	13.66	96.23	92.12	53.35	13.99	43.54	38.61	99.98	99.98	四 川
16.51	10.77	98.89	98.51	54.62	16.41	42.14	40.16	99.89	99.88	贵 州
14.71	12.50	99.02	97.48	65.26	13.94	43.14	39.52	100.00	99.87	云 南
9.97	3.23	96.57	96.57	27.18	16.23	40.80	38.60	99.81	99.81	西 藏
14.69	8.23	97.14	97.14	80.98	13.15	42.62	38.42	100.00	100.00	陕 西
15.23	8.84	97.79	97.79	74.52	16.36	36.20	33.08	100.00	100.00	甘 肃
15.48	14.33	95.89	95.89	64.16	13.24	36.53	34.22	99.46	99.46	青 海
15.78	4.49	99.04	99.04	76.49	22.84	42.15	40.40	100.00	100.00	宁 夏
12.57	6.72	97.79	97.78	82.02	14.93	41.17	38.01	100.00	100.00	新 疆
13.57	8.08	100.15	99.87	64.59	26.36	42.78	41.24	100.00	100.00	新疆生产建设兵团

Note: All of the green coverage rate and green space rate for the built-up areas of Beijing Municipality in the table refer to the data for the areas surveyed in the city. The green coverage rate and green space rate for the nationwide urban built-up areas have been revised appropriately.

1-2-1　全国历年城市数量及人口、面积情况

面积计量单位：平方公里
人口计量单位：万人

年 份 Year	城市个数 Number of Cities	地级 City at Prefecture Level	县级 City at County Level	县及其他 个　数 Number of Counties	城区人口 Urban Population
1978	193	98	92	2153	7682.0
1979	216	104	109	2153	8451.0
1980	223	107	113	2151	8940.5
1981	226	110	113	2144	14400.5
1982	245	109	133	2140	14281.6
1983	281	137	141	2091	15940.5
1984	300	148	149	2069	17969.1
1985	324	162	159	2046	20893.4
1986	353	166	184	2017	22906.2
1987	381	170	208	1986	25155.7
1988	434	183	248	1936	29545.2
1989	450	185	262	1919	31205.4
1990	467	185	279	1903	32530.2
1991	479	187	289	1894	29589.3
1992	517	191	323	1848	30748.2
1993	570	196	371	1795	33780.9
1994	622	206	413	1735	35833.9
1995	640	210	427	1716	37789.9
1996	666	218	445	1696	36234.5
1997	668	222	442	1693	36836.9
1998	668	227	437	1689	37411.8
1999	667	236	427	1682	37590.0
2000	663	259	400	1674	38823.7
2001	662	265	393	1660	35747.3
2002	660	275	381	1649	35219.6
2003	660	282	374	1642	33805.0
2004	661	283	374	1636	34147.4
2005	661	283	374	1636	35923.7
2006	656	283	369	1635	33288.7
2007	655	283	368	1635	33577.0
2008	655	283	368	1635	33471.1
2009	654	283	367	1636	34068.9
2010	657	283	370	1633	35373.5
2011	657	284	369	1627	35425.6
2012	657	285	368	1624	36989.7
2013	658	286	368	1613	37697.1
2014	653	288	361	1596	38576.5
2015	656	291	361	1568	39437.8
2016	657	293	360	1537	40299.2
2017	661	294	363	1526	40975.7
2018	673	302	371	1519	42730.0
2019	679	300	379	1516	43503.7
2020	687	301	386	1495	44253.7
2021	692	300	392	1482	45747.9
2022	695	302	393	1481	47001.9

注：1.2005年及以前年份"城区人口"为"城市人口"，"城区面积"为"城市面积"。
　　2.2005年、2009年、2011年城市建设用地面积不含上海市。2020年、2021年、2022年城区面积、建成区面积和城市建设用地面积不含北京市。

National Changes in Number of Cities, Urban Population and Urban Area in Past Years

Area Measurement Unit: Square Kilometer
Population Measurement Unit: 10,000 persons

非农业人口 Non-Agricultural Population	城区暂住人口 Urban Temporary Population	城区面积 Urban Area	建成区面积 Area of Built District	城市建设用地面积 Area of Urban Construction Land	年份 Year
					1978
					1979
					1980
9243.6		206684.0	7438.0	6720.0	1981
9590.0		335382.3	7862.1	7150.5	1982
10047.2		366315.9	8156.3	7365.6	1983
10956.9		480733.3	9249.0	8480.4	1984
11751.3		458066.2	9386.2	8578.6	1985
12233.8		805834.0	10127.3	9201.6	1986
12893.1		898208.0	10816.5	9787.9	1987
13969.5		1052374.2	12094.6	10821.6	1988
14377.7		1137643.5	12462.2	11170.7	1989
14752.1		1165970.0	12855.7	11608.3	1990
14921.0		980685.0	14011.1	12907.9	1991
15459.4		969728.0	14958.7	13918.1	1992
16550.1		1038910.0	16588.3	15429.8	1993
17665.5		1104712.0	17939.5	20796.2	1994
18490.0		1171698.0	19264.2	22064.0	1995
18882.9		987077.9	20214.2	19001.6	1996
19469.9		835771.8	20791.3	19504.6	1997
19861.8		813585.7	21379.6	20507.6	1998
20161.6		812817.6	21524.5	20877.0	1999
20952.5		878015.0	22439.3	22113.7	2000
21545.5		607644.3	24026.6	24192.7	2001
22021.2		467369.3	25972.6	26832.6	2002
22986.8		399173.2	28308.0	28971.9	2003
23635.9		394672.5	30406.2	30781.3	2004
23652.0		412819.1	32520.7	29636.8	2005
	3984.1	166533.5	33659.8	34166.7	2006
	3474.3	176065.5	35469.7	36351.7	2007
	3517.2	178110.3	36295.3	39140.5	2008
	3605.4	175463.6	38107.3	38726.9	2009
	4095.3	178691.7	40058.0	39758.4	2010
	5476.8	183618.0	43603.2	41805.3	2011
	5237.1	183039.4	45565.8	45750.7	2012
	5621.1	183416.1	47855.3	47108.5	2013
	5951.5	184098.6	49772.6	49982.7	2014
	6561.5	191775.5	52102.3	51584.1	2015
	7414.0	198178.6	54331.5	52761.3	2016
	8164.1	198357.2	56225.4	55155.5	2017
	8421.7	200896.5	58455.7	56075.9	2018
	8911.9	200569.5	60312.5	58307.7	2019
	9509.0	186628.9	60721.3	58355.3	2020
	10180.1	188300.5	62420.5	59424.6	2021
	9486.2	191216.8	63676.4	59451.7	2022

Notes: 1.In 2005 and before, Urban Population is the population of the city proper, and Urban Area is the area of the city proper.

2.Urban usable land for construction purpose throughout the country does not include that in Shanghai in 2005 ,2009 and 2011. Urban Area,Area of Built District and Urban usable land for construction purpose throughout the country does not include that in Beijing in 2020，2021and 2022.

1-2-2 全国城市人口和建设用地(2022年)

面积计量单位: 平方公里
人口计量单位: 万人

地区名称 Name of Regions	市区面积 Urban District Area	市区人口 Urban District Population	市区暂住人口 Urban District Temporary Population	城区面积 Urban Area	城区人口 Urban Population	城区暂住人口 Urban Temporary Population	建成区面积 Area of Built District	小计 Subtotal
全 国 National Total	**2371373**	**84820**	**13319**	**191216.77**	**47001.93**	**9486.24**	**63676.40**	**59451.69**
北 京 Beijing		2184			1912.80			
天 津 Tianjin	11926	1363		2653.44	1160.07		1264.46	1088.48
河 北 Hebei	49721	3793	211	6363.96	1816.65	187.78	2266.96	2225.36
山 西 Shanxi	36502	1707	251	3320.81	1097.11	182.96	1295.08	1260.05
内 蒙 古 Inner Mongolia	148683	904	320	4674.87	681.97	253.79	1272.80	1231.70
辽 宁 Liaoning	77574	3101	415	13067.76	2040.68	301.56	2815.03	2785.04
吉 林 Jilin	110793	1820	255	5720.76	970.92	228.71	1580.00	1499.02
黑 龙 江 Heilongjiang	220293	2159	212	2567.93	1198.93	177.75	1802.55	1619.35
上 海 Shanghai	6341	2476		6340.50	2475.89		1242.00	1093.36
江 苏 Jiangsu	70117	5912	1499	17190.16	3129.19	577.07	4916.17	4810.81
浙 江 Zhejiang	55798	3991	1800	13885.42	2186.38	1067.85	3426.99	3485.66
安 徽 Anhui	46910	2889	608	7125.58	1390.05	565.36	2500.28	2426.81
福 建 Fujian	48272	2299	827	4325.72	1108.23	402.34	1877.45	1605.45
江 西 Jiangxi	46475	2212	156	3337.68	1088.28	129.03	1789.49	1674.36
山 东 Shandong	93682	6838		24105.93	3513.75	642.96	5712.99	5480.54
河 南 Henan	50108	4990	568	6383.76	2386.06	473.72	3521.11	3353.56
湖 北 Hubei	97623	4258	720	7964.69	1938.93	495.05	2866.18	1915.79
湖 南 Hunan	56807	3199	386	4110.82	1697.15	241.81	2104.99	1935.99
广 东 Guangdong	99901	9842	1692	17126.58	5649.90	953.96	6575.31	6134.71
广 西 Guangxi	78641	2688	335	5394.85	1027.98	306.09	1809.48	1643.12
海 南 Hainan	16999	616	196	1340.03	232.66	100.57	419.28	375.19
重 庆 Chongqing	43264	2625	472	7781.33	1289.27	328.23	1640.80	1525.59
四 川 Sichuan	91215	4361	1050	8708.00	2369.47	826.65	3411.76	3198.44
贵 州 Guizhou	41821	1677	242	4355.99	708.30	203.16	1194.80	1064.14
云 南 Yunnan	91744	1834	194	3276.38	948.45	129.39	1287.83	1170.57
西 藏 Tibet	47488	94	38	632.58	56.36	39.56	170.71	164.57
陕 西 Shaanxi	60228	2146	219	2700.93	1270.49	166.75	1553.51	1513.35
甘 肃 Gansu	89185	909	194	2120.76	534.05	149.39	968.07	980.83
青 海 Qinghai	203426	295	27	738.91	192.28	21.45	250.08	232.73
宁 夏 Ningxia	21522	444	62	954.37	256.81	39.31	485.66	451.39
新 疆 Xinjiang	245051	1034	319	2187.14	597.22	259.11	1428.05	1294.74
新疆生产建设兵团 Xinjiang Production and Construction Corps	13262	158	51	759.13	75.65	34.88	226.53	210.99

注: 1.本表中山东省市区人口为常住人口。
　　2.部分省份"绿地与广场用地面积"根据第三次全国国土调查数据进行了校核修订。

National Urban Population and Construction Land (2022)

Area Measurement Unit: Square Kilometer
Population Measurement Unit: 10,000persons

居住用地 Residential	公共管理与公共服务用地 Administration and Public Services	商业服务业设施用地 Commercial and Business Facilities	工业用地 Industrial, Manufacturing	物流仓储用地 Logistics and Warehouse	道路交通设施用地 Road, Street and Transportation	公用设施用地 Municipal Utilities	绿地与广场用地 Green Space and Square	本年征用土地面积 Area of Land Requisition This Year	耕地 Arable Land	地区名称 Name of Regions
18823.58	5274.79	4298.80	11414.84	1534.15	10052.09	1355.04	6698.40	1946.70	823.43	全　国
										北　京
312.00	87.54	95.11	262.47	54.33	158.80	20.23	98.00	6.79	3.34	天　津
686.29	166.69	149.06	239.43	64.39	413.18	58.98	447.34	51.62	26.22	河　北
420.33	155.15	91.85	186.83	34.71	211.38	26.70	133.10	28.35	10.90	山　西
428.79	111.52	103.34	128.74	39.36	225.98	28.81	165.16	41.83	14.46	内　蒙　古
916.96	173.74	186.32	688.26	99.23	413.95	64.92	241.66	27.43	16.63	辽　宁
516.88	114.27	82.45	297.44	47.46	226.73	52.95	160.84	28.30	15.62	吉　林
551.64	133.77	93.37	380.88	55.82	217.70	30.84	155.33	22.89	16.63	黑　龙　江
375.59	102.32	107.01	177.91	42.99	190.82	14.96	81.76	17.24	8.81	上　海
1319.89	379.41	353.66	1073.76	91.34	856.59	102.21	633.95	149.31	83.44	江　苏
1108.01	323.45	242.49	771.29	56.37	615.91	63.96	304.18	109.68	57.93	浙　江
704.78	179.33	151.71	548.86	40.30	459.56	45.14	297.13	177.05	107.28	安　徽
579.70	154.63	117.41	282.85	33.71	256.22	28.69	152.24	60.33	15.86	福　建
540.04	164.21	100.95	304.47	27.34	291.71	30.63	215.01	76.82	28.40	江　西
1781.53	490.01	448.29	1232.23	149.16	789.81	98.88	490.63	195.30	49.79	山　东
1053.75	327.07	167.91	401.50	93.92	600.42	100.48	608.51	89.96	50.42	河　南
566.09	160.64	113.98	463.46	45.89	329.27	64.31	172.15	160.61	72.99	湖　北
745.22	215.27	134.61	290.33	38.43	265.50	62.91	183.72	40.07	14.07	湖　南
1946.23	495.97	427.55	1520.57	97.69	1213.68	98.26	334.76	123.35	32.39	广　东
491.33	143.47	108.28	237.36	49.99	304.44	54.42	253.83	107.35	36.49	广　西
151.85	51.48	54.39	18.05	10.37	54.22	6.26	28.57	21.26	3.39	海　南
464.97	141.33	85.95	333.46	36.84	308.96	29.56	124.52	66.68	23.66	重　庆
1022.96	293.16	240.40	553.93	86.95	535.34	71.04	394.66	149.40	62.41	四　川
363.85	133.27	103.38	157.87	31.10	153.59	22.81	98.27	38.25	17.23	贵　州
405.28	138.16	120.75	94.78	33.21	199.01	26.97	152.41	47.67	19.37	云　南
34.99	24.80	15.94	20.15	4.51	23.83	9.75	30.60	0.82	0.12	西　藏
444.90	129.97	146.39	241.84	47.93	245.61	35.58	221.13	38.07	16.01	陕　西
255.10	85.69	81.46	212.42	37.33	153.83	30.84	124.16	33.24	7.36	甘　肃
60.83	17.70	17.84	23.93	14.18	41.52	12.36	44.37	0.94		青　海
132.47	43.96	36.54	45.13	11.42	77.87	8.69	95.31	12.78	6.80	宁　夏
383.34	113.54	104.81	197.36	55.20	182.08	49.00	209.41	19.36	3.30	新　疆
57.99	23.27	15.60	27.28	2.68	34.58	3.90	45.69	3.95	2.11	新疆生产建设兵团

Notes: 1.The Urban District Population of Shandong Province in this table refers to the data of the resident population.

2.According to the data of the third National Land Survey,the Green Space and Square Land Area in some provinces has been checked and revised.

1-3-1 全国历年城市维护建设资金收入

计量单位: 万元

年份 Year	合计 Total	城市维护 建设税 Urban Maintenance and Construction Tax	城市公用 事业附加 Extra- Charges for Municipal Utilities	中央财政 拨款 Financial Allocation From Central Government Budget	地方财政 拨款 Financial Allocation From Local Government Budget
1978					
1979					
1980	276174		92966	66212	
1981	345600		92186	65657	
1982	425957		97416	69381	
1983	408524		103142	81571	
1984	468835		118445	96120	
1985	1168293	331296	129744	143313	123963
1986	1448311	406817	157889	138944	351961
1987	1638197	442275	176337	152094	225498
1988	1845193	525452	192643	104666	181053
1989	1835625	603583	198409	91529	170347
1990	2104896	650908	226186	109126	198421
1991	2661198	695681	270109	99919	277595
1992	3934788	779701	310916	134237	576569
1993	5811904	980234	329620	269593	594615
1994	6748008	1161193	413851	213565	599378
1995	7743732	1409097	447388	213543	725758
1996	8476420	1578088	555873	103699	862641
1997	11103424	1906696	561217	139392	1154245
1998	14333158	2153203	603965	649420	1581006
1999	16271209	2191985	631558	1052000	1714628
2000	19889324	2372908	541475	1150668	2081330
2001	25262680	2709430	488144	895818	3237790
2002	31561758	3160358	498761	759545	3927309
2003	42761892	3717417	556909	771320	5329302
2004	52575966	4462912	590251	526414	6657678
2005	54225147	5512933	554799	621597	7958871
2006	35406259	5667745	762171	566632	10748459
2007	47617452	6170604	824251	348134	12137926
2008	56164219	7442775	896248	756011	14247316
2009	67276878	7719472	980433	1066464	22311926
2010	85704996	9970340	1090649	1747961	16855095
2011	117817189	13395522	1571916	1383183	19817577
2012	119233467	14797543	1663517	2453300	27252469
2013	143227465	17070039	1874070		
2014	138005449	17843612	1819986		
2015	160735509	19421884	2255346		
2016	180194600	22429127	2100389		

注: 1.自2006年起, 城市维护建设资金收入仅包含财政性资金, 不含社会融资。地方财政拨款中包括省、市财政专项拨款和市级以下财政资金; 其他收入中包括市政公用设施配套费、市政公用设施有偿使用费、土地出让转让金、资产置换收入及其他财政性资金。

2.2014年数据不含北京市。

National Revenue of Urban Maintenance and Construction Fund in Past Years

Measurement Unit: 10,000 RMB

水资源费 Water Resource Fee	国内贷款 Domestic Loan	利用外资 Foreign Investment	企事业单位自筹资金 Self-Raised Funds by Enterprises and Institutions	其他收入 Other Revenues	年份 Year
					1978
					1979
				116996	1980
				187757	1981
				259160	1982
				223811	1983
				254270	1984
14142			49362	376473	1985
16812	31716	744		343428	1986
22420	61601	748		557224	1987
24396	75829	6386		734768	1988
24873	44494	10078		692312	1989
27876	88429	24713		779237	1990
35324	229274	107127		946169	1991
42506	322596	73484		1694779	1992
48498	445658	138159		3005527	1993
47060	413327	183295		3716339	1994
51964	476654	254876		4164452	1995
60509	956850	558688	1194722	2605348	1996
61203	1657405	1407194	1077986	3138088	1997
63724	3069622	738402	1769411	3704405	1998
77828	3741979	494463	2374802	3991966	1999
100590	4146988	847124	3332242	5315999	2000
111087	7416589	563167	4095561	5745094	2001
123825	8739016	610535	6007620	7734789	2002
159757	13318273	681057	7696801	10531056	2003
206883	14455465	741815	9002038	15932510	2004
249990	16698914	927141	9460261	12240641	2005
244522				17416730	2006
279545				27856992	2007
254199				32567671	2008
247971				34950612	2009
295347				55845604	2010
750347				80898644	2011
413782				72652856	2012
					2013
					2014
					2015
					2016

Notes: 1.Since 2006, national revenue of urban maintenance and construction fund includes the fund fiscal budget, not including social funds. Local Financail Allocation include province and city special financial allocation, and Other Revenues include fees for expansion of municipal utilities capacity, fees for use of municipal utilities, land transfer income, asset replacement income and other financial funds.

2.The data of 2014 did not include those of Beijing Municipality.

1-3-2　全国历年城市维护建设资金支出

计量单位：万元

年份 Year	支出合计 Total	按用途分　By Purpose			供水 Water Supply	燃气 Gas Supply	集中供热 Central Heating
		维护支出 Maintenance Expenditure	固定资产投资支出 Expenditure from Investment in Fixed Assets	其他支出 Other Expenditures			
1978	163589		163589		38439		
1979	256895	142593	114302		49317		
1980	266451	152991	112370	1090	45258		
1981	325615	180778	144837		32932		
1982	369382	172902	183897	12583	50238		
1983	426986	214709	185511	26766	51642		
1984	462705	219226	243479		61886		
1985	1122714	375563	721943	25208	102424	86598	12157
1986	1343025	431595	816154	95276	139303	110697	17993
1987	1480706	479768	895695	105243	50997	99993	20339
1988	644695	303420	246835	94440	78875		
1989	737038	361213	249516	126309	99822		
1990	814901	406025	269992	138884	108132		
1991	2442190	446562	1800536	195092	265064	224584	
1992	3841602	1086759	2529896	224947	453996	261810	
1993	5573279	1550715	3757865	264699	613695	394564	
1994	6583051	1721457	4514337	347257	753151	376545	
1995	7578470	1092891	6056988	428591	839579	332546	
1996	8827009	2029361	6272955	524693	991630	405386	149145
1997	11229697	2149827	7832004	1247866	915220	477144	255485
1998	14374965	2354278	10927777	1092910	1250356	552624	406733
1999	16172386	2600945	12281137	1290304	1175347	485483	425240
2000	18960338	2699820	15400602	859916	1232155	620982	638120
2001	25372970	2367551	19207735	3797684	1530995	775355	789525
2002	31780253	3184473	24475402	4120378	1578903	844230	1137775
2003	42479605	3840924	32807384	5831298	1769459	1120184	1438295
2004	46617124	4662527	36924771	5029826	2013230	1237042	1538473
2005	52757240	5474379	40946915	6335946	2140834	1210925	1838492
2006	33494964	6950298	20005661	6888341	791690	344728	583932
2007	42473049	8689485	24649406	9134158	981690	409311	617348
2008	50083394	10432730	30612349	9279334	1074319	565441	1047382
2009	59270667	10982852	38125424	10187001	1175896	512700	1206084
2010	75080799	13443275	47144531	14389306	1751365	823824	1450987
2011	87390666	16790270	53139075	17461321			
2012	101981275	20228860	63505657	18246758			
2013	108047393						
2014	106589135						
2015	124386269						
2016	138326496						

注：1.1978年至1984年，1988年至1990年供水支出为公用事业支出；1978年至2000年道路桥梁支出为市政工程支出，包含道路、桥梁、排水等。

2.自2006年起，城市维护建设资金支出仅包含财政性资金，不含社会融资。

3."公共交通"中，2009年至2010年数据仅包括城市轨道交通建设投资支出。

4.2014年数据不含北京市。

National Expenditure of Urban Maintenance and Construction Fund in Past Years

Measurement Unit: 10,000 RMB

公共交通 Public Transportation	道路桥梁 Road and Bridge	排　水 Sewerage	防　洪 Flood Control	园林绿化 Landscaping	市容环境 卫　　生 Environmental Sanitation	其　他 Other	年　份 Year
	48882			13904			1978
	67940			19337	26541		1979
	77891			24708	26093	1089	1980
	65032			23311	31751		1981
	114715			30304	41112	12585	1982
	121204			35445	44873	26766	1983
	150554			44421	52459	-6799	1984
56073	331930			85831	76250	25208	1985
43281	408526	86533		93234	107719	95276	1986
50997	481682	92141		98347	109356	105242	1987
	214471			61143	89510	94440	1988
	229082			66320	94189	126309	1989
	256539			73356	114880	138884	1990
69668	757003			186574	206776	195092	1991
106503	1198511			245334	283664	224947	1992
199547	2336633			311418	363825	264699	1993
139599	2929744			402105	440382	347257	1994
154127	3222256			523299	575786	428591	1995
201337	3590192	393753		532873	616247	524693	1996
285625	4664378	539945		675787	736913	1247866	1997
509899	6263259	881923		1035178	990036	1092909	1998
504331	7359954	985336		1315396	953377	1290304	1999
1434521	8188958	1291761		1779059	1526633	859916	2000
1901819	8640772	2181103	568835	1776331	932631	6275604	2001
2714424	11081731	2670957	1394590	918210	2628912	6810521	2002
2540364	17463295	3580353	1089181	3348090	1665167	8465217	2003
2405912	19862495	3361557	1494379	953154	3619398	10131484	2004
3467801	23719550	3555419	1786861	908127	4224763	9904468	2005
1511917	13154564	2215124	448841	2967691	1746169	9730308	2006
3394326	16951020	2972994	729391	3611468	2100669	10704832	2007
4012497	19769532	3709526	977739	4086639	2579521	12260798	2008
2259540	23061301	4923007	820261	4979200	2716274	17616404	2009
2530155	30396564	5808060	1604103	6908970	3666153	20140618	2010
							2011
							2012
							2013
							2014
							2015
							2016

Notes: 1.Data in the water supply column from 1978 to 1984 and from 1988 to 1990 refer to expenditure of public utilities; Data in the bridge and road column reder to expenditure of municipal engineering construction projects,including roads, bridges and sewers.

2.From 2006, expenditure of Urban maintenance and construction fund includes the fund from fiscal buget, not including society fund.

3.In the section of "Public Transit", data from 2009 to 2010 only included investment in the construction of urban rail infrastructure.

4.The data of 2014 did not include those of Beijing Municipality.

1-4-1 按行业分全国历年城市市政公用设施建设固定资产投资

计量单位: 亿元

年 份 Year	本 年 固 定 资 产 投资总额 Completed Investment of This Year	供 水 Water Supply	燃 气 Gas Supply	集中供热 Central Heating	轨道交通 Rail Transit System	道路桥梁 Road and Bridge	排 水 Sewerage
1978	12.0	4.7				2.9	
1979	14.2	3.4	0.6		1.8	3.1	1.2
1980	14.4	6.7				7.0	
1981	19.5	4.2	1.8		2.6	4.0	2.0
1982	27.2	5.6	2.0		3.1	5.4	2.8
1983	28.2	5.2	3.2		2.8	6.5	3.3
1984	41.7	6.3	4.8		4.7	12.2	4.3
1985	64.0	8.1	8.2		6.0	18.6	5.6
1986	80.1	14.3	12.5	1.6	5.6	20.5	6.0
1987	90.3	17.2	10.9	2.1	5.5	27.1	8.8
1988	113.2	23.1	11.2	2.8	6.0	35.6	10.0
1989	107.0	22.2	12.3	3.3	7.7	30.1	9.7
1990	121.2	24.8	19.4	4.5	9.1	31.3	9.6
1991	170.9	30.2	24.8	6.4	9.8	51.8	16.1
1992	283.2	47.7	25.9	11.0	14.9	90.6	20.9
1993	521.8	69.9	34.8	10.7	22.1	191.8	37.0
1994	666.0	90.3	32.5	13.4	25.1	279.8	38.3
1995	807.6	112.4	32.9	13.8	30.9	291.6	48.0
1996	948.6	126.1	48.3	15.7	38.8	354.2	66.8
1997	1142.7	128.3	76.0	25.1	43.2	432.4	90.1
1998	1477.6	161.0	82.0	37.3	86.1	616.2	154.5
1999	1590.8	146.7	72.1	53.6	103.1	660.1	142.0
2000	1890.7	142.4	70.9	67.8	155.7	737.7	149.3
2001	2351.9	169.4	75.5	82.0	194.9	856.4	224.5
2002	3123.2	170.9	88.4	121.4	293.8	1182.2	275.0
2003	4462.4	181.8	133.5	145.8	281.9	2041.4	375.2
2004	4762.2	225.1	148.3	173.4	328.5	2128.7	352.3
2005	5602.2	225.6	142.4	220.2	476.7	2543.2	368.0
2006	5765.1	205.1	155.0	223.6	604.0	2999.9	331.5
2007	6418.9	233.0	160.1	230.0	852.4	2989.0	410.0
2008	7368.2	295.4	163.5	269.7	1037.2	3584.1	496.0
2009	10641.5	368.8	182.2	368.7	1737.6	4950.6	729.8
2010	13363.9	426.8	290.8	433.2	1812.6	6695.7	901.6
2011	13934.2	431.8	331.4	437.6	1937.1	7079.1	770.1
2012	15296.4	410.4	414.5	630.3	2064.5	7402.5	704.5
2013	16350.0	524.7	425.6	596.0	2455.1	8355.6	778.9
2014	16245.0	475.3	416.0	575.4	3221.2	7643.9	900.0
2015	16204.4	619.9	350.5	516.8	3707.1	7414.0	982.7
2016	17460.0	545.8	408.9	481.9	4079.5	7564.3	1222.5
2017	19327.6	580.1	445.7	584.2	5045.2	6996.7	1343.6
2018	20123.2	543.0	295.1	420.0	6046.9	6922.4	1529.9
2019	20126.3	560.1	242.7	333.0	5855.6	7655.3	1562.4
2020	22283.9	749.4	238.6	393.8	6420.8	7814.3	2114.8
2021	23371.7	770.6	229.6	397.3	6339.0	8644.5	2078.8
2022	22309.9	713.3	286.0	339.8	6038.6	8707.9	1905.1

注: 1.2008年及以前年份，"轨道交通"投资为"公共交通"，2009年以后仅包含轨道交通建设投资。

2.自2013年开始，全国城市市政公用设施建设固定资产投资中不再包括城市防洪固定资产投资。

National Fixed Assets Investment in Urban Service Facilities by Industry in Past Years

Measurement Unit: 100 million RMB

污水处理及其再生利用 Wastewater Treatment and reused	防 洪 Flood Control	园 林 绿 化 Land-scaping	市容环境卫 生 Environmental Sanitation	垃圾处理 Garbage Treatment	地下综合管 廊 Utility Tunnel	其 他 Other	年 份 Year
						4.4	1978
	0.1	0.4	0.1			3.4	1979
						0.7	1980
	0.2	0.9	0.7			3.2	1981
	0.3	1.1	0.9			5.9	1982
	0.4	1.2	0.9			4.7	1983
	0.5	2.0	0.9			5.9	1984
	0.9	3.3	2.0			11.3	1985
	1.6	3.4	2.8			11.9	1986
	1.4	3.4	2.1			11.9	1987
	1.6	3.4	2.6			16.9	1988
	1.2	2.8	2.8			14.8	1989
	1.3	2.9	2.9			15.4	1990
	2.1	4.9	3.6			21.3	1991
	2.9	7.2	6.5			55.6	1992
	5.9	13.2	10.6			125.8	1993
	8.0	18.2	10.9			149.7	1994
	9.5	22.5	13.6			232.5	1995
	9.1	27.5	12.5			249.7	1996
	15.5	45.1	20.9			266.1	1997
	35.8	78.4	36.7			189.6	1998
	43.0	107.1	37.1			226.0	1999
	41.9	143.2	84.3			297.5	2000
116.4	70.5	163.2	50.6	23.5		466.6	2001
144.1	135.1	239.5	64.8	29.7		551.0	2002
198.8	124.5	321.9	96.0	35.3		760.4	2003
174.5	100.3	359.5	107.8	53.0		838.4	2004
191.4	120.0	411.3	147.8	56.7		947.0	2005
151.7	87.1	429.0	175.8	51.8		554.3	2006
212.2	141.4	525.6	141.8	53.0		735.6	2007
264.7	119.6	649.8	222.0	50.6		530.8	2008
418.6	148.6	914.9	316.5	84.6		923.9	2009
521.4	194.4	1355.1	301.6	127.4		952.2	2010
420.5	243.8	1546.2	384.1	199.2		773.1	2011
279.4	249.2	1798.7	296.5	110.9		1325.5	2012
353.7		1647.4	408.4	125.9		1158.0	2013
404.2		1817.6	494.8	130.6		700.9	2014
512.6		1594.7	398.0	157.0		620.7	2015
489.9		1670.1	445.2	118.1	294.7	747.0	2016
450.8		1759.6	508.1	240.8	673.4	1391.0	2017
802.6		1854.7	470.5	298.5	619.2	1421.4	2018
803.7		1844.8	557.4	406.8	558.1	956.9	2019
1043.4		1626.3	862.6	705.8	453.6	1609.6	2020
893.8		1638.6	727.1	535.9	538.9	2007.3	2021
708.2		1347.6	485.0	304.1	307.6	2179.0	2022

Notes: 1. For the year 2008 and before, there was data about investment in public transport. Starting from 2009, the data has only included investment in rail transport construction.

2. Starting from 2013, the national fixed assets investment in the construction of municipal public utilities facilities has not included the fixed assets investment in urban flood prevention.

1-4-2　按行业分全国城市市政公用设施建设固定资产投资(2022年)

计量单位: 万元

地区名称 Name of Regions		本年完成 投　资 Completed Investment of This Year	供　水 Water Supply	燃　气 Gas Supply	集中供热 Central Heating	轨道交通 Urban Rail Trainsit System	道路桥梁 Road and Bridge	地下综合 管　廊 Utility Tunnel
全　国	**National Total**	**223098515**	**7133438**	**2860066**	**3398079**	**60385551**	**87079061**	**3075548**
北　京	Beijing	12901497	410295	133714	272665	3378087	3337959	77109
天　津	Tianjin	4590015	110613	61425	64774	3413906	355301	16946
河　北	Hebei	6585015	302393	101425	512682	108813	2760598	608764
山　西	Shanxi	3909165	57159	92567	578548	319595	1919180	76475
内　蒙　古	Inner Mongolia	1907580	93525	66363	250037	3828	831134	650
辽　宁	Liaoning	3877368	139246	190552	162604	1895979	925457	37061
吉　林	Jilin	3649915	114366	74398	91787	1176723	659863	39537
黑　龙　江	Heilongjiang	2366508	537258	120872	181561	275019	653510	
上　海	Shanghai	5863849	148730	81492		2803600	1380917	27385
江　苏	Jiangsu	19414966	497418	302548	17662	5860690	8461208	254606
浙　江	Zhejiang	18667886	335805	106161		5968205	8462365	157917
安　徽	Anhui	11052577	438730	199719	27972	1706839	5613522	104940
福　建	Fujian	6687039	243028	79525		2525707	2129263	227669
江　西	Jiangxi	6636730	190507	54475		481211	4042718	108878
山　东	Shandong	14885880	382455	281999	613579	2846445	6364971	232862
河　南	Henan	7336304	179946	61678	157371	2164302	1930158	69830
湖　北	Hubei	15290359	234532	82282	51087	3753155	6348893	278012
湖　南	Hunan	7194373	264743	94743	4300	1546812	3061519	1828
广　东	Guangdong	18196943	915349	285643		7846541	5864268	244772
广　西	Guangxi	3007068	266646	33873		206545	1744917	27702
海　南	Hainan	1086875	3856	236			805831	2445
重　庆	Chongqing	11737858	189315	29052		2624382	5940589	31420
四　川	Sichuan	16693378	363118	130213	8030	4996093	7410973	75061
贵　州	Guizhou	4633156	280578	91697	4444	945364	1104678	1200
云　南	Yunnan	1709372	58934	4628		17962	426561	87358
西　藏	Tibet	179033	9748	1600	24236		91844	16504
陕　西	Shaanxi	8995630	156640	68703	124933	3101898	3186904	77078
甘　肃	Gansu	1703529	50069	14745	51930	334600	601793	35767
青　海	Qinghai	264058	14933	4936	31145		81300	10710
宁　夏	Ningxia	295532	66206	2548	26316		89664	
新　疆	Xinjiang	1416787	60929	5442	101197	83250	351037	127324
新疆生产 建设兵团	Xinjiang Production and Construction Corps	362270	16368	812	39219		140166	17738

National Fixed Assets Investment in Urban Service Facilities by Industry(2022)

Measurement Unit: 10,000RMB

排水 Sewerage	污水处理 Wastewater Treatment	污泥处置 Sludge Disposal	再生水利用 Wastewater Recycled and Reused	园林绿化 Landscaping	市容环境卫生 Environmental Sanitation	垃圾处理 Domestic Garbage Treatment	其他 Other	本年新增固定资产 Newly Added Fixed Assets of This Year	地区名称 Name of Regions
19050915	6729410	567537	352543	13476470	4849535	3041162	21789852	86593149	全 国
589126	57015	4651	22295	886889	219745	20271	3595908	3737170	北 京
172831	93578		938	87110	45184	44010	261925	3348533	天 津
800476	126871	479	23087	913309	256272	133489	220283	1704572	河 北
333319	128614	12423	10500	196218	113794	85734	222310	1172394	山 西
184063	37198	26016	34380	162714	63803	57109	251463	303221	内 蒙 古
125482	37762	4335	647	43522	178496	170754	178969	450898	辽 宁
265205	223845	1795	3123	163024	8336	6386	1056676	599703	吉 林
264973	64039	10500		67646	51344	49957	214325	391838	黑 龙 江
699266	283705	200039		443304	61918	50118	217237	801949	上 海
1969966	1049398	16916	16721	1385643	233168	117660	432057	4649145	江 苏
914504	433594	11856	5292	1196641	131931	32223	1394357	10590341	浙 江
1212977	264140	16883	7205	589149	385351	111267	773378	2274949	安 徽
575332	228149	6778	8400	324722	137805	111411	443988	1426907	福 建
698557	287550	5197		529479	91037	40864	439868	3434688	江 西
1620614	289555	28735	6419	1018797	221880	127248	1302278	6517627	山 东
742006	212426	46614	34955	1287894	406726	391813	336393	7142509	河 南
827747	132407	9378	6662	593944	434987	366555	2685720	3597233	湖 北
589135	128629	69138	25002	80375	85732	42206	1465186	1509918	湖 南
1700533	829799	410	17390	166616	419773	374724	753448	8496242	广 东
360141	142693	5	11570	162693	152713	81956	51838	833083	广 西
116810	48992	5742	25238	118865	5996	1946	32836	8605	海 南
603241	144593	6771	3803	1023428	150810	56616	1145621	8769236	重 庆
1792608	722134	47477	3855	1116418	592840	336959	208024	9532853	四 川
220070	134661			5524	76152	64676	1903449	757296	贵 州
685200	161109	6507	23113	83487	26482	17176	318760	1173894	云 南
23150	8616			2283	6914	6914	2754	148771	西 藏
542498	231051	26050	30490	540383	184969	61133	1011624	1182112	陕 西
161505	120145		6677	110933	31564	18244	310623	1076073	甘 肃
79308	39757	1809		12790	563	126	28373	115242	青 海
72368	2166	1033	11010	17136	17030	17030	4264	59906	宁 夏
100297	65016		13771	112262	50144	40647	424905	401825	新 疆
7607	203			33272	6076	3940	101012	384416	新疆生产建设兵团

1-5-1 按资金来源分全国历年城市市政公用设施建设固定资产投资

计量单位：亿元

年 份 Year	本年资金 来源合计 Completed Investment of This Year	上 年 末 结余资金 The Balance of The Previous Year	本年资金来源		
			小 计 Subtotal	中央财政 拨 款 Financial Allocation From Central Government Budget	地方财政 拨 款 Financial Allocation From Local Government Budget
1978	12.0		6.4	6.4	
1979	14.0				
1980	14.4		14.4	6.1	
1981	20.0		20.0	5.3	
1982	27.2		27.2	8.6	
1983	28.2		28.2	8.2	
1984	41.7		41.7	11.8	
1985	63.8		63.8	13.9	
1986	79.8		79.8	13.2	
1987	90.0		90.0	13.4	
1988	112.6		112.6	10.2	
1989	106.8		106.8	9.7	
1990	121.2		121.2	7.4	
1991	169.9		169.9	8.6	
1992	265.4		265.4	9.9	
1993	521.6		521.6	15.9	
1994	665.5		665.5	27.9	
1995	837.0		807.5	24.2	
1996	939.1	68.0	871.1	34.8	
1997	1105.6	49.0	1056.6	43.0	
1998	1404.4	58.0	1346.4	100.2	
1999	1534.2	81.0	1453.2	173.8	
2000	1849.5	109.0	1740.5	222.0	
2001	2351.9	109.0	2112.8	104.9	379.1
2002	3123.2	111.0	2705.9	96.3	516.9
2003	4264.1	120.7	4143.4	118.9	733.4
2004	4650.9	267.9	4383.0	63.0	938.4
2005	5505.5	229.0	5276.6	63.9	1050.6
2006	5800.6	365.4	5435.2	89.2	1339.0
2007	6283.5	369.5	5914.0	77.3	1925.7
2008	7277.4	386.9	6890.4	72.7	2143.9
2009	10938.1	460.4	10477.6	112.9	2705.1
2010	13351.7	659.3	12692.4	206.0	3523.6
2011	14158.1	648.9	13509.1	166.3	4555.6
2012	15264.2	595.4	14668.9	171.1	4446.6
2013	16121.9	987.5	15134.3	147.5	3573.2
2014	16054.0	954.0	15100.0	102.2	4135.2
2015	16570.7	1295.0	15275.8	202.1	4406.4
2016	17319.2	942.7	16376.5	119.4	5183.7
2017	19704.7	1459.6	18245.1	290.4	5465.7
2018	19084.8	1245.5	17839.3	255.2	4552.3
2019	20438.8	1568.1	18870.7	412.6	5456.9
2020	23265.7	2228.6	21037.1	568.6	5922.0
2021	25778.0	3656.4	22121.6	661.6	6122.8
2022	22062.3	1932.1	20130.2	301.5	5853.3

注：自2013年起，"本年资金来源合计"为"本年实际到位资金合计"。

National Fixed Assets Investment in Urban Service Facilities by Capital Source in Past Years

Measurement Unit: 100 million RMB

| Sources of Fund | | | | | |
国内贷款 Domestic Loan	债　券 Securities	利用外资 Foreign Investment	自筹资金 Self- Raised Funds	其他资金 Other Funds	年　份 Year
					1978
					1979
0.1			8.2		1980
0.4			13.8	0.5	1981
1.0			16.5	1.1	1982
0.6			17.2	2.2	1983
2.1			25.5	2.3	1984
3.1		0.1	40.9	5.8	1985
3.1		0.1	57.4	6.0	1986
6.2		1.3	61.4	7.7	1987
7.1		1.6	78.0	15.7	1988
6.0		1.5	72.9	16.7	1989
11.0		2.2	82.2	18.4	1990
25.3		6.0	108.1	21.9	1991
42.9		10.3	180.4	38.9	1992
72.8		20.8	304.5	107.6	1993
58.3		64.2	397.1	118.0	1994
65.1		84.9	493.3	140.0	1995
122.2	4.9	105.6	486.1	117.5	1996
165.3	3.1	129.5	554.3	161.4	1997
284.8	40.3	110.1	600.4	210.6	1998
357.8	55.9	68.6	595.2	201.9	1999
428.6	29.0	76.7	682.7	301.5	2000
603.4	16.8	97.8	636.4	274.5	2001
743.8	7.3	109.6	866.3	365.7	2002
1435.4	17.4	90.0	1350.2	398.0	2003
1468.0	8.5	87.2	1372.9	445.0	2004
1805.9	5.2	170.0	1728.0	453.0	2005
1880.5	16.4	92.9	1638.1	379.2	2006
1763.7	29.5	73.1	1635.7	409.0	2007
2037.0	27.8	91.2	1980.1	537.6	2008
4034.8	120.8	66.1	2487.1	950.7	2009
4615.6	49.1	113.8	3058.9	1125.3	2010
3992.8	111.6	100.3	3478.6	1103.9	2011
4366.7	26.8	150.8	3740.5	1766.4	2012
4218.0	41.5	62.2	4714.1	2377.7	2013
4383.1	96.0	42.0	4294.7	2046.7	2014
3986.3	189.1	46.6	4258.0	2187.3	2015
4338.7	133.4	34.6	3963.6	2603.0	2016
4987.5	163.2	28.7	4997.8	2311.7	2017
4509.0	117.2	46.9	5105.2	3253.5	2018
5039.3	392.2	49.9	4707.3	2812.5	2019
3932.6	1864.0	67.5	5343.3	3339.1	2020
3566.2	1428.1	39.0	5789.1	4514.8	2021
2855.5	1723.4	35.6	5321.8	4039.2	2022

Note: Since 2013, Completed Investment of This Year is changed to be The Total Funds Actually Available for The Reported Year.

1-5-2　按资金来源分全国城市市政公用设施 建设固定资产投资(2022年)

计量单位: 万元

地区名称 Name of Regions	本年实际到位资金合计 The Total Funds Actually Available for The Reported Year	上年末结余资金 The Balance of The Previous Year	本年资金来源			
			小计 Subtotal	国家预算资金 State Budgetary Fund	中央预算资金 Central Budgetary Fund	国内贷款 Domestic Loan
全　国　National Total	220622915	19320735	201302180	61547819	3015261	28555412
北　京　Beijing	12268215	1371173	10897042	4169123	392635	1531187
天　津　Tianjin	4522051	658996	3863055	1105561	14905	1611760
河　北　Hebei	7258230	1204329	6053901	2449446	443270	329100
山　西　Shanxi	3233976	77089	3156887	1081995	44008	302157
内　蒙　古　Inner Mongolia	1397994	92349	1305645	511476	27248	29057
辽　宁　Liaoning	2454361	175430	2278931	492565	106741	547620
吉　林　Jilin	4122528	632990	3489538	422322	170988	1221107
黑　龙　江　Heilongjiang	2380756	223239	2157517	1407277	241774	178439
上　海　Shanghai	5377062	163962	5213100	2175365		31567
江　苏　Jiangsu	20900425	2283263	18617162	4200948	8775	2584045
浙　江　Zhejiang	20048800	2458545	17590255	4038142	92232	3859495
安　徽　Anhui	11940288	233087	11707201	4884178	130540	931292
福　建　Fujian	6573094	1036487	5536607	2365868	83325	193698
江　西　Jiangxi	7633184	673211	6959973	2710319	2924	490326
山　东　Shandong	14190096	799790	13390306	3097633	21854	2296639
河　南　Henan	7416281	540856	6875425	2293989	184291	1783079
湖　北　Hubei	10316498	835376	9481122	2608254	229283	453267
湖　南　Hunan	6409964	313872	6096092	380864	25114	56705
广　东　Guangdong	19447841	619023	18828818	6149208	90761	3593502
广　西　Guangxi	2591399	125803	2465596	943886	21086	439976
海　南　Hainan	1079284	54885	1024399	895474	21800	6132
重　庆　Chongqing	13613223	631278	12981945	5761977	87253	2848420
四　川　Sichuan	15548966	362758	15186208	4640835	212127	1668033
贵　州　Guizhou	3646294	363993	3282301	110242	10768	276667
云　南　Yunnan	1132066	312577	819489	234004	21821	35656
西　藏　Tibet	606297	32646	573651	48465	9172	
陕　西　Shaanxi	10077386	2445434	7631952	1744286	119570	733485
甘　肃　Gansu	2106268	359693	1746575	60928	45918	265703
青　海　Qinghai	266596	61142	205454	145069	23802	348
宁　夏　Ningxia	332083	17017	315066	53038	7865	143232
新　疆　Xinjiang	1228265	83631	1144634	156065	39329	113718
新疆生产建设兵团　Xinjiang Production and Construction Corps	503144	76811	426333	209017	84082	

National Fixed Assets Investment in Urban Service Facilities by Capital Source(2022)

Measurement Unit: 10,000 RMB

Sources of Fund					
债券 Securities	利用外资 Foreign Investment	自筹资金 Self-Raised Funds	其他资金 Other Funds	各项应付款 Sum Payable This Year	地区名称 Name of Regions
17233528	355557	53217677	40392187	31457204	全　国
		2485200	2711532	987869	北　京
211118	16041	851028	67547	549372	天　津
2058849		836891	379615	2106377	河　北
254299		640684	877752	716767	山　西
195427	4099	278182	287404	879648	内　蒙　古
572265	6294	487574	172613	1009975	辽　宁
829603	4628	426912	584966	702717	吉　林
193887		218629	159285	274633	黑　龙　江
		3006168		680347	上　海
344511	35180	9499647	1952831	1620570	江　苏
493393	45012	6537252	2616961	1630249	浙　江
1712955	42986	1674391	2461399	1386716	安　徽
1165516	19	968496	843010	627258	福　建
158161		2634899	966268	2235755	江　西
1809850	20417	4099884	2065883	2743028	山　东
597074	71460	1604470	525353	791462	河　南
876829	1237	1819419	3722116	966201	湖　北
174716	9600	3946259	1527948	226255	湖　南
1536495		1458136	6091477	987701	广　东
171483	14284	737048	158919	724690	广　西
77936		9509	35348	153833	海　南
1205058		2030666	1135824	1652939	重　庆
971422	6500	1985178	5914240	3564481	四　川
32830		1397532	1465030	946484	贵　州
121838		165825	262166	909547	云　南
178681		48954	297551		西　藏
281589		2603553	2269039	1158830	陕　西
242616	77800	632703	466825	604458	甘　肃
42119		9337	8581	1173	青　海
45644		46732	26420	41622	宁　夏
519772		74859	280220	555431	新　疆
157592		1660	58064	20816	新疆生产建设兵团

二、居民生活数据

Data by Residents Living

1-6-1　全国历年城市供水情况

年　份 Year	综　合 生产能力 （万立方米/日） Integrated Production Capacity (10,000 m³/day)	供水管道 长　度 （公里） Length of Water Supply Pipelines (km)	供水总量 （万立方米） Total Quantity of Water Supply (10,000 m³)	生活用量 Residential Use
1978	2530.4	35984	787507	275854
1979	2714.0	39406	832201	309206
1980	2979.0	42859	883427	339130
1981	3258.0	46966	969943	367823
1982	3424.9	51513	1011319	391422
1983	3539.0	56852	1065956	421968
1984	3960.9	62892	1176474	465651
1985	4019.7	67350	1280238	519493
1986	10407.9	72557	2773921	706971
1987	11363.6	77864	2984697	759702
1988	12715.8	86231	3385847	873800
1989	12821.1	92281	3936648	930619
1990	14220.3	97183	3823425	1001021
1991	14584.0	102299	4085073	1159929
1992	16036.4	111780	4298437	1172919
1993	16927.9	123007	4502341	1282543
1994	18215.1	131052	4894620	1422453
1995	19250.4	138701	4815653	1581451
1996	19990.0	202613	4660652	1670673
1997	20565.8	215587	4767788	1757157
1998	20991.8	225361	4704732	1810355
1999	21551.9	238001	4675076	1896225
2000	21842.0	254561	4689838	1999960
2001	22900.0	289338	4661194	2036492
2002	23546.0	312605	4664574	2131919
2003	23967.1	333289	4752548	2246665
2004	24753.0	358410	4902755	2334625
2005	24719.8	379332	5020601	2437374
2006	26965.6	430426	5405246	2220459
2007	25708.4	447229	5019488	2263676
2008	26604.1	480084	5000762	2274266
2009	27046.8	510399	4967467	2334082
2010	27601.5	539778	5078745	2371488
2011	26668.7	573774	5134222	2476520
2012	27177.3	591872	5230326	2572473
2013	28373.4	646413	5373022	2676463
2014	28673.3	676727	5466613	2756911
2015	29678.3	710206	5604728	2872695
2016	30320.7	756623	5806911	3031376
2017	30475.0	797355	5937591	3153968
2018	31211.8	865017	6146244	3300567
2019	30897.8	920082	6283010	3401160
2020	32072.7	1006910	6295420	3484644
2021	31737.7	1059901	6733442	3753783
2022	31510.4	1102976	6744063	3785485

注：1.1978年至1985年综合供水生产能力为系统内数；1978年至1995年供水管道长度为系统内数。

　　2.自2006年起，供水普及率指标按城区人口和城区暂住人口合计为分母计算，括号中的数据为往年同口径数据。

National Urban Water Supply in Past Years

用水人口 （万人） Population with Access to Water Supply (10,000 persons)	人　均　日 生活用水量 （升） Daily Water Consumption Per Capita (liter)	供　水 普及率 (%) Water Coverage Rate (%)	年　份 Year
6267.1	120.6	81.6	1978
6951.0	121.8	82.3	1979
7278.0	127.6	81.4	1980
7729.3	130.4	53.7	1981
8102.2	132.4	56.7	1982
8370.9	138.1	52.5	1983
8900.7	143.3	49.5	1984
9424.3	151.0	45.1	1985
11757.9	161.9	51.3	1986
12684.6	164.1	50.4	1987
14049.9	170.4	47.6	1988
14786.3	172.4	47.4	1989
15611.1	175.7	48.0	1990
16213.2	196.0	54.8	1991
17280.8	186.0	56.2	1992
18636.4	188.6	55.2	1993
20083.0	194.0	56.0	1994
22165.7	195.4	58.7	1995
21997.0	208.1	60.7	1996
22550.1	213.5	61.2	1997
23169.1	214.1	61.9	1998
23885.7	217.5	63.5	1999
24809.2	220.2	63.9	2000
25832.8	216.0	72.26	2001
27419.9	213.0	77.85	2002
29124.5	210.9	86.15	2003
30339.7	210.8	88.85	2004
32723.4	204.1	91.09	2005
32304.1	188.3	86.67(97.04)	2006
34766.5	178.4	93.83	2007
35086.7	178.2	94.73	2008
36214.2	176.6	96.12	2009
38156.7	171.4	96.68	2010
39691.3	170.9	97.04	2011
41026.5	171.8	97.16	2012
42261.4	173.5	97.56	2013
43476.3	173.7	97.64	2014
45112.6	174.5	98.07	2015
46958.4	176.9	98.42	2016
48303.5	178.9	98.30	2017
50310.6	179.7	98.36	2018
51778.0	180.0	98.78	2019
53217.4	179.4	98.99	2020
55580.9	185.0	99.38	2021
56141.8	184.7	99.39	2022

Notes: 1.Integrated production capacity from 1978 to 1985 is limited to the statistical figure in building sector; Length of water supply pipelines from 1978 to 1995 is limited to the statistical figure in building sector.

2.Since 2006,water coverage rate has been calculated based on denominator which combines both permanent and temporary residents in urban areas, and the data in brackets are the same index but calculated by the method of past years.

1-6-2　城市供水(2022年)

地区名称 Name of Regions			综合 生产能力 (万立方米/日) Integrated Production Capacity (10,000m³/ day)	地下水 Underground Water	供水管道 长度 (公里) Length of Water Supply Pipelines (km)	建成区 in Built District	供水总量 (万立方米) Total Quantity of Water Supply (10,000m³)	生产运营 用水 The Quantity of Water for Production and Operation
全　　国		**National Total**	**31510.40**	**3610.01**	**1102975.66**	**958648.60**	**6744062.93**	**1679461.83**
北　京		Beijing	702.71	254.23	19553.00	11675.13	149875.61	11734.87
天　津		Tianjin	505.60	15.00	22574.16	21396.64	102041.50	34872.55
河　北		Hebei	788.50	276.14	23616.16	22745.16	156165.22	40200.93
山　西		Shanxi	397.47	201.44	16213.99	14083.98	91584.27	16994.67
内　蒙　古		Inner Mongolia	450.46	274.22	13515.32	12722.74	82147.86	22676.67
辽　宁		Liaoning	1364.26	284.46	40735.70	38019.15	277004.58	75896.04
吉　林		Jilin	657.23	85.70	16705.49	16145.57	102363.56	23628.22
黑　龙　江		Heilongjiang	595.34	200.32	24939.27	24438.00	129184.52	31642.60
上　海		Shanghai	1229.00		40017.93	40017.93	292326.62	42017.57
江　苏		Jiangsu	3653.28	65.72	130938.06	93068.21	643682.45	223838.90
浙　江		Zhejiang	2131.67	0.23	105237.85	75062.67	474877.77	158823.40
安　徽		Anhui	1143.61	90.50	35915.72	34847.92	261030.95	80573.26
福　建		Fujian	958.58	13.31	34397.20	33086.87	198962.35	31983.67
江　西		Jiangxi	715.69	2.77	31992.33	30028.51	161381.69	30532.87
山　东		Shandong	1986.52	514.58	61342.56	58057.00	399552.84	154652.25
河　南		Henan	1322.52	331.21	31893.30	30771.88	234881.38	50559.16
湖　北		Hubei	1610.48	5.35	56312.48	54387.42	340856.76	91045.08
湖　南		Hunan	1056.63	18.44	41224.94	38002.83	248659.84	44224.17
广　东		Guangdong	3979.22	23.08	150032.98	120516.29	1010915.61	251150.71
广　西		Guangxi	776.28	33.94	26593.44	25800.98	200674.71	35847.27
海　南		Hainan	206.98	18.27	8127.88	3901.16	50048.45	3602.59
重　庆		Chongqing	794.46	1.80	26937.77	24339.41	183613.48	38598.15
四　川		Sichuan	1367.98	86.24	58292.64	56022.70	339871.37	50794.68
贵　州		Guizhou	474.54	1.30	24378.71	21879.22	95932.41	17675.97
云　南		Yunnan	514.04	20.35	18491.40	17031.80	112538.67	17747.99
西　藏		Tibet	68.70	31.90	1883.65	1859.95	14370.71	1449.63
陕　西		Shaanxi	645.77	208.39	12794.77	12319.90	140284.77	36268.59
甘　肃		Gansu	374.47	56.77	7247.28	6165.92	58843.68	15396.91
青　海		Qinghai	139.25	110.36	3584.19	3248.80	32280.14	13287.02
宁　夏		Ningxia	275.12	87.82	3503.96	3350.96	38321.16	9131.20
新　疆		Xinjiang	539.35	289.13	11936.94	11706.37	97256.35	15239.22
新疆生产 建设兵团		Xinjiang Production and Construction Corps	84.69	7.04	2044.59	1947.53	22531.65	7375.02

Urban Water Supply(2022)

公共服务 用　水 The Quantity of Water for Public Service	居民家庭 用　水 The Quantity of Water for Household Use	其他用水 The Quantity of Other Purposes	用水户数 （户） Number of Households with Access to Water Supply (unit)	家庭用户 Household User	用水人口 （万人） Population with Access to Water Supply (10,000 persons)	地区名称 Name of Regions
981163.54	2793442.54	308571.81	221105246	199201695	56141.81	全　　国
45389.29	68347.93	2850.36	7098424	6885640	1909.21	北　京
13446.44	38553.17	372.72	5198665	5015541	1160.06	天　津
21604.07	68437.70	4044.90	6771763	6301530	2004.43	河　北
15148.64	47767.32	2853.20	2995298	2848530	1262.86	山　西
12447.50	28159.33	5250.96	6094400	5426507	932.92	内 蒙 古
37710.37	91781.11	8353.43	13596407	12579014	2318.66	辽　宁
15114.18	36261.39	3014.12	6400132	5804921	1157.13	吉　林
18009.51	45230.48	8040.86	7693415	6987464	1364.35	黑 龙 江
68404.22	118696.44	9752.91	9548047	9030968	2475.89	上　海
72310.55	212710.23	53498.92	19639604	17672830	3706.26	江　苏
66232.73	188992.63	10775.53	12326836	11169591	3254.23	浙　江
36160.97	102098.64	9454.03	9424569	8292284	1950.71	安　徽
38525.93	87598.66	10885.97	5753488	5102997	1510.19	福　建
23083.64	74521.48	6281.02	6432303	5808618	1209.62	江　西
49976.87	141170.67	13534.79	13041599	12316085	4153.47	山　东
32389.67	112174.76	8587.95	8680746	7951374	2839.86	河　南
26750.18	150330.27	9588.91	8801734	8198695	2432.27	湖　北
37690.49	112472.61	12686.78	7331950	6575550	1919.71	湖　南
164765.69	415373.08	44845.41	21134978	16443859	6586.57	广　东
30650.44	102384.59	2825.95	3551372	3213387	1332.88	广　西
5749.64	29119.58	4860.15	384705	332770	333.07	海　南
25700.60	79154.63	12317.45	7977509	7383998	1594.44	重　庆
51102.57	169278.00	18810.21	12357967	10790247	3105.96	四　川
7524.20	50255.82	2478.10	3940151	3533299	901.29	贵　州
17867.54	54537.71	3697.08	4196379	3714369	1067.14	云　南
1911.86	6488.96	712.62	244153	191041	95.63	西　藏
9576.26	74157.48	4713.04	3833919	3661381	1412.12	陕　西
8652.55	25732.56	4233.56	1435358	1358010	680.04	甘　肃
3466.90	9885.01	2250.55	846794	759732	212.81	青　海
7600.33	10318.38	5902.47	1243359	1111536	296.09	宁　夏
12431.34	37053.82	17271.19	2729338	2377788	853.75	新　疆
3768.37	4398.10	3826.67	399884	362139	108.19	新疆生产 建设兵团

1-6-3 城市供水(公共供水)(2022年)

地区名称 Name of Regions	综合生产能力 (万立方米/日) Integrated Production Capacity (10,000 m³/day)	地下水 Underground Water	水厂个数 (个) Number of Water Plants (unit)	地下水 Underground Water	供水管道长度 (公里) Length of Water Supply Pipelines (km)	供水总量(万立方米) 合计 Total	售水量 小计 Subtotal	生产运营用水 The Quantity of Water for Production and Operation
全 国 National Total	28452.31	2666.96	3019	820	1081956.80	6354506.37	5373083.16	1362361.00
北 京 Beijing	657.52	209.04	68	45	18997.57	144820.25	123267.09	11082.73
天 津 Tianjin	475.00	15.00	31	7	22517.74	98912.58	84115.96	31743.63
河 北 Hebei	730.92	237.03	138	76	21508.26	140858.47	118980.85	27315.45
山 西 Shanxi	368.22	174.19	79	60	15633.16	87234.11	78413.67	14738.91
内 蒙 古 Inner Mongolia	364.04	205.59	82	77	13239.90	68139.43	54526.03	13709.62
辽 宁 Liaoning	1160.70	213.14	159	63	39952.82	246356.46	183092.83	50049.38
吉 林 Jilin	380.89	45.50	69	22	16266.16	85301.62	60955.97	8880.42
黑 龙 江 Heilongjiang	531.79	149.91	96	56	24220.91	116757.05	90495.98	22868.79
上 海 Shanghai	1229.00		39		40017.93	292326.62	238871.14	42017.57
江 苏 Jiangsu	3028.70	36.00	128	14	129929.73	583945.49	502621.64	165407.96
浙 江 Zhejiang	2029.19		130		102667.94	458880.52	408827.04	143406.58
安 徽 Anhui	919.22	64.72	76	10	34522.21	226272.76	193528.71	51066.17
福 建 Fujian	942.17	10.90	90	9	34305.44	197725.71	167757.59	31018.98
江 西 Jiangxi	701.70	1.50	79		31737.12	159539.16	132576.48	29322.47
山 东 Shandong	1755.59	372.41	294	110	59187.39	346626.44	306408.18	108881.38
河 南 Henan	1122.34	192.19	149	65	30442.51	206723.68	175553.84	32855.77
湖 北 Hubei	1453.69		117		55328.95	315679.07	252536.75	68746.50
湖 南 Hunan	1048.30	14.40	94	9	39498.54	246354.89	204769.10	43056.79
广 东 Guangdong	3785.18	19.00	249	6	149768.67	1007205.15	872424.43	247496.12
广 西 Guangxi	728.13	29.00	83	7	25398.08	190295.59	161329.13	26142.90
海 南 Hainan	202.50	14.25	19	1	7927.88	49173.50	42457.01	2970.79
重 庆 Chongqing	733.47		96		26857.57	180275.86	152433.21	35891.67
四 川 Sichuan	1276.25	34.95	169	15	58196.70	334536.06	284650.15	46478.80
贵 州 Guizhou	474.36	1.14	92	3	24374.91	95919.75	77921.43	17674.76
云 南 Yunnan	479.13	12.50	118	11	18372.47	106861.15	88172.80	13510.46
西 藏 Tibet	64.72	27.92	19	11	1883.65	14310.48	10502.84	1389.40
陕 西 Shaanxi	548.45	144.61	89	47	12012.23	124703.42	109134.02	29068.91
甘 肃 Gansu	349.96	52.26	43	21	7077.51	57464.68	52636.58	14382.35
青 海 Qinghai	80.66	63.66	13	9	3195.32	19577.69	16187.03	1729.45
宁 夏 Ningxia	233.88	57.38	26	16	3121.96	35008.67	29639.89	7639.21
新 疆 Xinjiang	511.95	261.73	61	35	11750.98	94188.41	78927.63	14442.06
新疆生产建设兵团 Xinjiang Production and Construction Corps	84.69	7.04	24	15	2044.59	22531.65	19368.16	7375.02

Urban Water Supply(Public Water Suppliers)(2022)

Total Quantity of Water Supply(10,000m³)						用水户数 (户) Number of Households with Access to Water Supply (unit)	居民家庭 Households	用水人口 (万人) Population with Access to Water Supply (10,000 persons)	地区名称 Name of Regions
Water Sold			免费供水量 The Quantity of Free Water Supply	生活用水 Domestic Water Use	漏损水量 The Lossed Water				
公共服务用水 The Quantity of Water for Public Service	居民家庭用水 The Quantity of Water for Household Use	其他用水 The Quantity of Other Purposes							
951369.33	2762596.10	296756.73	162369.72	10878.70	819053.49	219520553	197909461	55769.89	全　　国
43042.16	66352.91	2789.29	198.45	5.00	21354.71	7012784	6811877	1839.21	北　　京
13446.44	38553.17	372.72	1244.22		13552.40	5198662	5015541	1160.06	天　　津
21252.39	66445.75	3967.26	3127.56	175.21	18750.06	6724337	6258187	1996.56	河　　北
15021.30	46004.73	2648.73	463.08	12.97	8357.36	2798451	2652554	1247.98	山　　西
9237.60	26773.17	4805.64	3011.40		10602.00	6074025	5410673	929.38	内 蒙 古
35614.62	90385.38	7043.45	22133.29	2375.33	41130.34	13345651	12422625	2288.02	辽　　宁
14146.30	35627.97	2301.28	4639.42	81.20	19706.23	6353141	5759633	1143.57	吉　　林
16703.39	43497.65	7426.15	3739.35	142.78	22521.72	7629567	6926869	1353.00	黑 龙 江
68404.22	118696.44	9752.91	8791.05		44664.43	9548047	9030968	2475.89	上　　海
71668.88	212185.39	53359.41	11063.35	1144.42	70260.50	19617786	17670640	3704.34	江　　苏
66114.31	188530.62	10775.53	3609.03	206.72	46444.45	12292090	11136043	3244.32	浙　　江
32450.64	100804.71	9207.19	5949.77	91.42	26794.28	9354749	8232859	1937.88	安　　徽
38525.93	87326.71	10885.97	7315.54	40.84	22652.58	5747302	5101914	1509.94	福　　建
23051.56	74039.79	6162.66	3171.89	622.99	23790.79	6412864	5792611	1207.27	江　　西
47014.43	138474.82	12037.55	3007.98	209.41	37210.28	12856231	12165153	4115.40	山　　东
27326.01	109255.31	6116.75	5467.88	440.93	25701.96	8507670	7797887	2780.56	河　　南
26154.69	148509.01	9126.55	14184.97	409.55	48957.35	8691704	8140863	2421.15	湖　　北
37120.59	111932.71	12659.01	10713.99	1938.45	30871.80	7326136	6572523	1914.67	湖　　南
164719.82	415368.08	44840.41	14123.73	519.73	120656.99	21126837	16435859	6584.17	广　　东
30587.07	101845.65	2753.51	4919.19	134.97	24047.27	3537989	3200488	1327.20	广　　西
5654.40	28974.66	4857.16	1168.60	25.99	5547.89	384598	332769	327.20	海　　南
25673.55	78704.68	12163.31	2075.81	344.74	25766.84	7938368	7348578	1581.59	重　　庆
50396.44	169033.22	18741.69	7578.93	1042.24	42306.98	12345213	10783305	3103.66	四　　川
7520.78	50255.01	2470.88	2875.16	53.91	15123.16	3939701	3532864	901.17	贵　　州
17867.54	53477.96	3316.84	5940.64	273.86	12747.71	4195385	3713671	1063.88	云　　南
1911.86	6488.96	712.62	2317.09	174.09	1490.55	244153	191041	95.63	西　　藏
6789.60	69682.08	3593.43	3159.80	24.87	12409.60	3704461	3539369	1374.76	陕　　西
8539.65	25523.36	4191.22	1036.38	86.60	3791.72	1419816	1343031	675.81	甘　　肃
3465.53	8763.50	2228.55	1393.56	233.79	1997.10	829356	742780	210.28	青　　海
6622.04	9965.29	5413.35	1301.05		4067.73	1242376	1110836	295.84	宁　　夏
11557.22	36719.31	16209.04	2493.80	26.69	12766.98	2721219	2373311	851.31	新　　疆
3768.37	4398.10	3826.67	153.76	40.00	3009.73	399884	362139	108.19	新疆生产建设兵团

1-6-4 城市供水(自建设施供水)(2022年)

地区名称 Name of Regions	综合生产能力 (万立方米/日) Integrated Production Capacity (10,000 m³/day)	地下水 Underground Water	供水管道长度 (公里) Length of Water Supply Pipelines (km)	建成区 in Built District	供水总量(万立方米)	
					合计 Total	生产运营用水 The Quantity of Water for Production and Operation
全 国 National Total	**3058.09**	**943.05**	**21018.86**	**13981.02**	**389556.56**	**317100.83**
北 京 Beijing	45.19	45.19	555.43		5055.36	652.14
天 津 Tianjin	30.60		56.42	56.42	3128.92	3128.92
河 北 Hebei	57.58	39.11	2107.90	1825.70	15306.75	12885.48
山 西 Shanxi	29.25	27.25	580.83	135.41	4350.16	2255.76
内 蒙 古 Inner Mongolia	86.42	68.63	275.42	173.42	14008.43	8967.05
辽 宁 Liaoning	203.56	71.32	782.88	256.38	30648.12	25846.66
吉 林 Jilin	276.34	40.20	439.33	292.83	17061.94	14747.80
黑 龙 江 Heilongjiang	63.55	50.41	718.36	405.60	12427.47	8773.81
上 海 Shanghai						
江 苏 Jiangsu	624.58	29.72	1008.33	906.41	59736.96	58430.94
浙 江 Zhejiang	102.48	0.23	2569.91	810.61	15997.25	15416.82
安 徽 Anhui	224.39	25.78	1393.51	1305.75	34758.19	29507.09
福 建 Fujian	16.41	2.41	91.76	91.76	1236.64	964.69
江 西 Jiangxi	13.99	1.27	255.21	230.89	1842.53	1210.40
山 东 Shandong	230.93	142.17	2155.17	1843.77	52926.40	45770.87
河 南 Henan	200.18	139.02	1450.79	1048.00	28157.70	17703.39
湖 北 Hubei	156.79	5.35	983.53	636.66	25177.69	22298.58
湖 南 Hunan	8.33	4.04	1726.40	1103.60	2304.95	1167.38
广 东 Guangdong	194.04	4.08	264.31	18.58	3710.46	3654.59
广 西 Guangxi	48.15	4.94	1195.36	1120.36	10379.12	9704.37
海 南 Hainan	4.48	4.02	200.00		874.95	631.80
重 庆 Chongqing	60.99	1.80	80.20	72.50	3337.62	2706.48
四 川 Sichuan	91.73	51.29	95.94	27.80	5335.31	4315.88
贵 州 Guizhou	0.18	0.16	3.80	3.80	12.66	1.21
云 南 Yunnan	34.91	7.85	118.93	97.93	5677.52	4237.53
西 藏 Tibet	3.98	3.98			60.23	60.23
陕 西 Shaanxi	97.32	63.78	782.54	725.04	15581.35	7199.68
甘 肃 Gansu	24.51	4.51	169.77	132.57	1379.00	1014.56
青 海 Qinghai	58.59	46.70	388.87	147.27	12702.45	11557.57
宁 夏 Ningxia	41.24	30.44	382.00	326.00	3312.49	1491.99
新 疆 Xinjiang	27.40	27.40	185.96	185.96	3067.94	797.16
新疆生产建设兵团 Xinjiang Production and Construction Corps						

Urban Water Supply (Suppliers with Self-Built Facilities)(2022)

Total Quantity of Water Supply(10,000 m³)			用水户数（户）	居民家庭	用水人口（万人）	地区名称
公共服务用水 The Quantity of Water for Public Service	居民家庭用水 The Quantity of Water for Household Use	其他用水 The Quantity of Water for Other Purposes	Number of Households with Access to Water Supply (unit)	Households	Population with Access to Water Supply (10,000 persons)	Name of Regions
29794.21	30846.44	11815.08	1584693	1292234	371.92	全　国
2347.13	1995.02	61.07	85640	73763	70.00	北　京
			3			天　津
351.68	1991.95	77.64	47426	43343	7.87	河　北
127.34	1762.59	204.47	196847	195976	14.88	山　西
3209.90	1386.16	445.32	20375	15834	3.54	内　蒙　古
2095.75	1395.73	1309.98	250756	156389	30.64	辽　宁
967.88	633.42	712.84	46991	45288	13.56	吉　林
1306.12	1732.83	614.71	63848	60595	11.35	黑　龙　江
						上　海
641.67	524.84	139.51	21818	2190	1.92	江　苏
118.42	462.01		34746	33548	9.91	浙　江
3710.33	1293.93	246.84	69820	59425	12.83	安　徽
	271.95		6186	1083	0.25	福　建
32.08	481.69	118.36	19439	16007	2.35	江　西
2962.44	2695.85	1497.24	185368	150932	38.07	山　东
5063.66	2919.45	2471.20	173076	153487	59.30	河　南
595.49	1821.26	462.36	110030	57832	11.12	湖　北
569.90	539.90	27.77	5814	3027	5.04	湖　南
45.87	5.00	5.00	8141	8000	2.40	广　东
63.37	538.94	72.44	13383	12899	5.68	广　西
95.24	144.92	2.99	107	1	5.87	海　南
27.05	449.95	154.14	39141	35420	12.85	重　庆
706.13	244.78	68.52	12754	6942	2.30	四　川
3.42	0.81	7.22	450	435	0.12	贵　州
	1059.75	380.24	994	698	3.26	云　南
						西　藏
2786.66	4475.40	1119.61	129458	122012	37.36	陕　西
112.90	209.20	42.34	15542	14979	4.23	甘　肃
1.37	1121.51	22.00	17438	16952	2.53	青　海
978.29	353.09	489.12	983	700	0.25	宁　夏
874.12	334.51	1062.15	8119	4477	2.44	新　疆
						新疆生产建设兵团

1-7-1　全国历年城市节约用水情况

计量单位：万立方米

年　份 Year	计划用水量 Planned Quantity of Water Use	新水取用量 Fresh Water Used
1991	1977024	1921307
1992	2077092	1939390
1993	2076928	2004922
1994	2299596	2497236
1995	2123215	1930610
1996	6285246	2166806
1997	2004412	1875974
1998	2490390	2343370
1999	2409254	2223035
2000	2153979	2071731
2001	2504134	2126401
2002	2463717	2091365
2003	2353286	2014528
2004	2441914	2048488
2005	2676425	2300332
2006		2356268
2007		2295192
2008		2121034
2009		2064014
2010		2161262
2011		1694727
2012		1838306
2013		1876372
2014		1950107
2015		1922051
2016		1971043
2017		2455307
2018		2290261
2019		2333077
2020		2350149
2021		2559406
2022		2943120

注：自2006年起，不统计计划用水量指标。

National Urban Water Conservation in Past Years

Measurement Unit: 10,000 m³

工业用水 重复利用量 Quantity of Industrial Water Recycled	节约用水量 Water Saved	年 份 Year
1985326	211259	1991
2843918	205087	1992
3000109	218317	1993
3375652	276241	1994
3626086	235113	1995
3345351	235194	1996
4429175	260885	1997
3978967	278835	1998
3966355	287284	1999
3811945	353569	2000
3881683	377733	2001
4849164	372352	2002
4572711	338758	2003
4183937	393426	2004
5670096	376093	2005
5535492	415755	2006
6026048	454794	2007
6547258	659114	2008
6130645	628692	2009
6872928	407152	2010
6334101	406578	2011
7388185	400806	2012
6526517	382760	2013
6930588	405234	2014
7160130	403133	2015
7644277	576220	2016
8211738	648227	2017
8556264	508437	2018
9081103	499888	2019
10305444	707572	2020
10359838	738645	2021
11611625	707498	2022

Note: From 2006, "Planned Quantity of Water Use" has not been counted.

1-7-2 城市节约用水（2022年）

计量单位：万立方米

地区名称 Name of Regions	计划用水户数（户） Planned Water Consumers (unit)	自备水计划用水户数 Planned Self-produced Water Consumers	计划用水户实际用水量				
			合计 Total	工业 Industry	新水取水量 Fresh Water Used	工业 Industry	重复利用量 Water Reused
全 国 National Total	2072679	72643	15221173	12829221	2943120	1217595	12278053
北 京 Beijing	45823	3990	751355	530050	234301	17389	517055
天 津 Tianjin	12850	1162	1168379	1154427	41827	27892	1126552
河 北 Hebei	41021	674	261714	240100	34911	15471	226802
山 西 Shanxi	5796	228	459275	397416	82807	24879	376468
内 蒙 古 Inner Mongolia	426981	406	221418	201871	40905	21844	180513
辽 宁 Liaoning	68156	174	648154	591539	74433	35067	573720
吉 林 Jilin	4548	669	155674	109385	94296	48271	61378
黑 龙 江 Heilongjiang	1177	943	358448	242876	181547	66139	176901
上 海 Shanghai	173187		99980	39161	99980	39161	
江 苏 Jiangsu	29508	1270	2585732	2067238	347252	191856	2238480
浙 江 Zhejiang	43912	873	995569	861792	197478	121723	798091
安 徽 Anhui	6993	123	740724	704689	92048	58144	648677
福 建 Fujian	113703		221672	180962	61350	24380	160323
江 西 Jiangxi	5727	262	66636	33528	45923	20283	20713
山 东 Shandong	68764	4366	1603793	1358983	230187	115237	1373606
河 南 Henan	38185	8928	729247	687357	74032	39106	655215
湖 北 Hubei	143495	136	841424	777075	97977	58150	743447
湖 南 Hunan	88484	313	166696	136197	54297	26267	112400
广 东 Guangdong	199376	8362	1550682	1244506	427872	132314	1122810
广 西 Guangxi	8400	127	460173	397135	76614	18059	383560
海 南 Hainan	2933	107	57176	2881	57084	2880	92
重 庆 Chongqing	2483	65	46106	32342	24380	11939	21726
四 川 Sichuan	28946	4403	242683	187346	80725	40331	161958
贵 州 Guizhou	6944	195	82679	44962	43659	9557	39020
云 南 Yunnan	152961	532	109707	94338	26149	11160	83558
西 藏 Tibet	355		1416		1416		
陕 西 Shaanxi	38320	32165	52964	23314	41039	12585	11925
甘 肃 Gansu	256560	49	264503	239953	37514	14618	226989
青 海 Qinghai	5	2	2404	1015	924	421	1480
宁 夏 Ningxia	870	236	248508	241742	17003	10237	231505
新 疆 Xinjiang	8116	1883	24966	5034	21878	2225	3088
新疆生产建设兵团 Xinjiang Production and Construction Corps	48100		1313	10	1313	10	

Urban Water Conservation (2022)

Measurement Unit: 10,000m³

Actual Quantity of Water Used			工业 Industry	节约用水量 Water Saved	工业 Industry	节水措施投资总额（万元） Total Investment in Water-Saving Measures (10,000 RMB)	地区名称 Name of Regions
工业 Industry	超计划定额用水量 Water Quantity Consumed in Excess of Quota	重复利用率 (%) Reuse Rate (%)					
11611625	20771	81	91	707498	475208	748204	全　国
512661	229	69	97	7416	479	6579	北　京
1126535	1141	96	98	660	654	4839	天　津
224629	33	87	94	5641	3798	1219	河　北
372537	659	82	94	14779	7891	2065	山　西
180027	717	82	89	471	40	65792	内　蒙　古
556473	413	89	94	15989	8144	3178	辽　宁
61113	34	39	56	8100	5889	259	吉　林
176737	194	49	73	73589	22287	213	黑　龙　江
				2198	561	21583	上　海
1875382	1956	87	91	77985	57407	30928	江　苏
740069	981	80	86	41579	27545	66359	浙　江
646545	754	88	92	30419	27490	49586	安　徽
156582	2725	72	87	21226	12776	10958	福　建
13244	1600	31	40	8002	4102	3762	江　西
1243746	1247	86	92	49004	36123	136068	山　东
648251	1396	90	94	26573	15538	2750	河　南
718924	606	88	93	52782	48195	23212	湖　北
109930	722	67	81	111970	108719	15616	湖　南
1112193	2432	72	89	69896	32807	131233	广　东
379076	635	83	95	13942	9173	3370	广　西
1	267	0	0	233	37	47850	海　南
20403	32	47	63	9257	7854	61346	重　庆
147014	267	67	78	19707	9931	10916	四　川
35404	633	47	79	17820	17418	4515	贵　州
83177	535	76	88	4491	176	9449	云　南
							西　藏
10729	133	23	46	6166	3357	16665	陕　西
225335	35	86	94	7106	2405	12638	甘　肃
594	75	62	59	310	310	112	青　海
231505	159	93	96	3614	2164	5140	宁　夏
2809	161	12	56	6574	1938	5	新　疆
							新疆生产建设兵团

1-8-1　全国历年城市燃气情况
National Urban Gas in Past Years

年 份 Year	人工煤气　Man-Made Coal Gas				天然气　Natural Gas			
	供气总量 （万立方米） Total Gas Supplied (10,000 m³)	居民家庭 Households	用气人口 （万人） Population with Access to Gas (10,000 persons)	管道长度 （公里） Length of Gas Supply Pipeline (km)	供气总量 （万立方米） Total Gas Supplied (10,000 m³)	居民家庭 Households	用气人口 （万人） Population with Access to Gas (10,000 persons)	管道长度 （公里） Length of Gas Supply Pipeline (km)
1978	172541	66593	450	4157	69078	4103	24	560
1979	182748	73557	500	4446	68489	9833	76	751
1980	195491	83281	561	4698	58937	4893	80	921
1981	199466	90326	594	4830	93970	8575	83	1059
1982	208819	94896	630	5258	90421	8938	93	1098
1983	214009	89012	703	5967	49071	9443	91	1149
1984	231351	96228	768	7353	168568	38258	135	1965
1985	244754	107060	911	8255	162099	43268	281	2312
1986	337745	139374	951	7990	435887	65262	492	2409
1987	686518	170887	1177	11650	501093	77719	634	5465
1988	667927	171376	1357	13028	573983	73305	748	6186
1989	841297	204709	1498	14448	591169	91770	896	6849
1990	1747065	274127	1674	16312	642289	115662	972	7316
1991	1258088	311430	1867	18181	754616	154302	1072	8054
1992	1495531	305325	2181	20931	628914	152013	1119	8487
1993	1304130	345180	2490	23952	637207	139297	1180	8889
1994	1262712	416453	2889	27716	752436	146938	1273	9566
1995	1266894	456585	3253	33890	673354	163788	1349	10110
1996	1348076	472904	3490	38486	637832	138018	1470	18752
1997	1268944	535412	3735	41475	663001	177121	1656	22203
1998	1675571	480734	3746	42725	688255	195778	1908	25429
1999	1320925	494001	3918	45856	800556	215006	2225	29510
2000	1523615	630937	3944	48384	821476	247580	2581	33655
2001	1369144	494191	4349	50114	995197	247543	3127	39556
2002	1989196	490258	4541	53383	1259334	350479	3686	47652
2003	2020883	583884	4792	57017	1416415	374986	4320	57845
2004	2137225	512026	4654	56419	1693364	454248	5628	71411
2005	2558343	458538	4369	51404	2104951	521389	7104	92043
2006	2964500	381518	4067	50524	2447742	573441	8319	121498
2007	3223512	373522	4022	48630	3086365	662198	10190	155271
2008	3558287	353162	3370	45172	3680393	779917	12167	184084
2009	3615507	307134	2971	40447	4050996	913386	14544	218778
2010	2799380	268764	2802	38877	4875808	1171596	17021	256429
2011	847256	238876	2676	37100	6787997	1301190	19028	298972
2012	769686	215069	2442	33538	7950377	1558311	21208	342752
2013	627989	167886	1943	30467	8882417	1726620	23783	388466
2014	559513	145773	1757	29043	9643783	1968878	25973	434571
2015	471378	108306	1322	21292	10407906	2080061	28561	498087
2016	440944	108716	1085	18513	11717186	2864124	30856	551031
2017	270882	73733	752	11716	12637546	2825027	33934	623253
2018	297893	78957	779	13124	14439538	3135097	36902	698043
2019	276841	56168	675	10915	15279409	3470004	39025	767946
2020	231447	52031	548	9860	15637020	3815984	41302	850552
2021	187234	42792	456	9165	17210612	4119858	44196	929088
2022	181450	35207	381	6718	17677007	4382046	45679	980405

注：自2006年起，燃气普及率指标按城区人口和城区暂住人口合计为分母计算，括号中的数据为与往年同口径数据。

Note: Since 2006, gas coverage rate has been calculated based on denominator which combines both permanent and temporary residents in urban areas, and the data in brackets are the same index but calculated by the method of past years.

1-8-1 续表 continued

年 份 Year	液化石油气 LPG				燃气普及率 (%) Gas Coverage Rate (%)
	供气总量 (吨) Total Gas Supplied (ton)	居民家庭 Households	用气人口 (万人) Population with Access to Gas (10,000 persons)	管道长度 (公里) Length of Gas Supply Pipeline (km)	
1978	194533	175744	635		14.4
1979	243576	218797	788		16.1
1980	290460	269502	924		17.3
1981	330987	308028	995		11.6
1982	388596	342828	1076		12.6
1983	456192	414635	1159		12.3
1984	535289	424318	1435		13.0
1985	601803	540761	1534		13.0
1986	1011308	763590	2041		15.2
1987	1049116	954534	2399		16.7
1988	1730261	1136883	2764		16.5
1989	1898003	1259349	3156		17.8
1990	2190334	1428058	3579		19.1
1991	2423988	1694399	4084		23.7
1992	2996699	2019620	4796		26.3
1993	3150296	2316129	5770		27.9
1994	3664948	2817702	6745		30.4
1995	4886528	3701504	8355		34.3
1996	5758374	3943604	8864	2762	38.2
1997	5786023	4370979	9350	4086	40.0
1998	7972947	5478535	9995	4458	41.8
1999	7612684	4990363	10336	6116	43.8
2000	10537147	5322828	11107	7419	45.4
2001	9818313	5583497	13875	10809	60.42
2002	11363884	6561738	15431	12788	67.17
2003	11263475	7817094	16834	15349	76.74
2004	11267120	7041351	17559	20119	81.53
2005	12220141	7065214	18013	18662	82.08
2006	12636613	6936513	17100	17469	79.11(88.58)
2007	14667692	7280415	18172	17202	87.40
2008	13291072	6292713	17632	28590	89.55
2009	13400303	6887600	16924	14236	91.41
2010	12680054	6338523	16503	13374	92.04
2011	11658326	6329164	16094	12893	92.41
2012	11148032	6081312	15683	12651	93.15
2013	11097298	6130639	15102	13437	94.25
2014	10828490	5862125	14378	10986	94.57
2015	10392169	5871062	13955	9009	95.30
2016	10788042	5739456	13744	8716	95.75
2017	9988088	5447739	12616	6200	96.26
2018	10153298	5447936	11782	4841	96.70
2019	9227179	4917008	11297	4452	97.29
2020	8337109	4786679	10767	4010	97.87
2021	8606841	4936380	10180	2910	98.04
2022	7584586	4444194	9333	2547	98.06

1-8-2 城市人工煤气(2022年)

地区名称 Name of Regions	生产能力 (万立方米／日) Production Capacity (10,000 m³/day)	储气能力 (万立方米) Gas Storage Capacity (10,000 m³)	供气管道长度 (公里) Length of Gas Supply Pipeline (km)	自制气量 (万立方米) Self-Produced Gas (10,000 m³)	合计 Total
全　国　National Total	**780.66**	**132.86**	**6717.90**	**186443.42**	**181449.71**
河　北　Hebei			426.80		58144.01
山　西　Shanxi	225.00	19.00	659.60	150.00	51225.27
内　蒙　古　Inner Mongolia			195.93		2812.29
辽　宁　Liaoning	51.06	38.00	2584.00	11518.92	25653.42
吉　林　Jilin		8.60	349.00		3569.25
黑　龙　江　Heilongjiang		8.00	212.60		2033.00
江　西　Jiangxi	3.00	20.00	85.20	780.00	15249.04
山　东　Shandong	132.00	1.00	2.90	48180.00	7679.00
河　南　Henan	15.00		235.43		
湖　北　Hubei			589.11		
广　西　Guangxi	9.60	18.20	514.67		5193.68
四　川　Sichuan		13.00	811.66		7925.90
甘　肃　Gansu	345.00	7.00	51.00	125814.50	1960.80
新　疆　Xinjiang					
新疆生产建设兵团　Xinjiang Production and Construction Corps		0.06			4.05

Urban Man-Made Coal Gas(2022)

供气总量(万立方米) Total Gas Supplied (10,000 m³)		燃气损失量 Loss Amount	用气户数 (户) Number of Households with Access to Gas (unit)	居民家庭 Households	用气人口 (万人) Population with Access to Gas (10,000 persons)	地区名称 Name of Regions
销售气量 Quantity Sold	居民家庭 Households					
173564.52	35207.36	7885.19	1796019	1779791	380.51	**全　国**
57083.47		1060.54	104			河　北
50341.56	1128.95	883.71	59894	59778	15.20	山　西
2780.08	1928.21	32.21	65359	64972	12.21	内　蒙　古
22677.63	17394.34	2975.79	1016955	1006724	187.97	辽　宁
3472.21	2307.54	97.04	189000	187965	45.20	吉　林
1949.00	1686.00	84.00	106000	104000	30.61	黑　龙　江
14508.70	474.18	740.34	12365	12305	0.92	江　西
7678.00		1.00	1			山　东
						河　南
						湖　北
5071.19	3919.10	122.49	120271	119455	34.88	广　西
6155.48	4711.69	1770.42	159632	158503	37.32	四　川
1843.15	1653.30	117.65	65935	65586	14.00	甘　肃
						新　疆
4.05	4.05		503	503	2.20	新疆生产 建设兵团

1-8-3 城市天然气（2022年）

地区名称 Name of Regions	储气能力 （万立方米） Gas Storage Capacity (10,000 m³)	供气管道 长 度 （公里） Length of Gas Supply Pipeline (km)	供气总量(万立方米)			
			合 计 Total	销售气量 Quantity Sold	居民家庭 Households	集中供热 Central Heating
全 国 National Total	190702.71	980404.73	17677006.77	17364407.89	4382045.79	1753637.43
北 京 Beijing	19798.32	31881.26	1991095.00	1946384.00	204687.00	634911.00
天 津 Tianjin	721.77	52607.69	681389.09	670400.00	116069.60	188110.10
河 北 Hebei	8183.55	45247.85	633611.78	619331.13	225351.05	91537.76
山 西 Shanxi	2017.27	29297.96	320965.82	315063.17	95737.52	9009.89
内 蒙 古 Inner Mongolia	414.98	12564.89	255319.53	249041.15	96673.45	37154.27
辽 宁 Liaoning	2999.62	33291.54	380778.21	374103.43	85898.29	9637.61
吉 林 Jilin	7085.56	14247.47	228098.41	223702.41	47797.49	8250.67
黑 龙 江 Heilongjiang	870.23	12898.86	176772.26	171176.61	43547.09	26763.32
上 海 Shanghai	72900.00	33665.42	911453.62	887227.64	194652.16	
江 苏 Jiangsu	15980.88	114704.37	1717890.20	1691474.68	343601.17	7233.03
浙 江 Zhejiang	2638.84	63946.49	997007.03	988556.16	139776.20	25203.97
安 徽 Anhui	4530.38	35768.81	501513.81	488464.60	151168.85	60.00
福 建 Fujian	1030.06	18186.11	335961.66	334196.83	35039.96	
江 西 Jiangxi	1017.17	21979.54	251368.98	248496.72	74400.25	
山 东 Shandong	5099.81	83079.72	1275880.96	1254258.49	324671.46	105115.50
河 南 Henan	1048.83	30838.78	741158.33	724106.00	248569.51	72295.57
湖 北 Hubei	6199.52	51333.71	632308.01	615988.83	190433.70	302.87
湖 南 Hunan	3102.83	32744.47	332266.42	323935.88	147882.73	
广 东 Guangdong	2665.22	47463.50	1393326.76	1383272.55	201331.71	
广 西 Guangxi	1041.36	11469.23	206966.96	205751.41	55524.04	
海 南 Hainan	188.70	5538.65	35151.93	34590.27	21009.14	
重 庆 Chongqing	465.17	25933.22	591129.89	579064.39	234302.98	2289.00
四 川 Sichuan	555.25	77456.14	1028009.57	1004121.38	458435.64	252.58
贵 州 Guizhou	1184.58	10765.72	180290.08	178690.93	55730.45	370.70
云 南 Yunnan	337.66	10100.70	73686.84	72992.81	21268.69	
西 藏 Tibet	10.00	6165.85	5200.00	5100.00	2000.00	3100.00
陕 西 Shaanxi	9107.62	29677.91	617803.88	605701.62	229683.17	162093.84
甘 肃 Gansu	225.61	4658.72	264163.10	263323.11	69342.95	72259.59
青 海 Qinghai	235.63	4148.87	171968.32	166643.77	47619.88	67072.95
宁 夏 Ningxia	6550.64	7517.20	126664.83	125369.64	44985.40	13073.23
新 疆 Xinjiang	12211.05	18961.27	564173.82	560917.84	158932.26	213861.30
新疆生产建设兵团 Xinjiang Production and Construction Corps	284.60	2262.81	53631.67	52960.44	15922.00	3678.68

Urban Natural Gas(2022)

Total Gas Supplied (10,000 m³)		用气户数 (户) Number of Household with Access to Gas (unit)	居民家庭 Households	用气人口 (万人) Population with Access to Gas (10,000 persons)	天 然 气 汽车加气站 (座) Gas Stations for CNG-Fueled Motor Vehicles (unit)	地区名称 Name of Regions
燃气汽车 Gas-Powered Automobiles	燃气损失量 Loss Amount					
988145.12	312598.88	203986365	199606207	45678.51	4230	全　　国
23060.00	44711.00	7630618	7531949	1472.90	82	北　　京
26959.75	10989.09	6268591	5909873	1087.09	73	天　　津
44738.51	14280.65	9136545	8815983	1836.41	234	河　　北
22173.58	5902.65	5411163	5357444	1202.27	93	山　　西
36238.83	6278.38	3239450	3174661	705.27	191	内　蒙　古
30086.58	6674.78	9075596	8994821	1837.95	185	辽　　宁
34368.48	4396.00	4452858	4381584	902.55	242	吉　　林
33504.07	5595.65	4707048	4646139	1000.35	200	黑　龙　江
11205.00	24225.98	7990403	7850466	1981.24	19	上　　海
40192.35	26415.52	16233998	15605191	3382.99	225	江　　苏
27912.26	8450.87	8535595	8466247	2281.45	110	浙　　江
35013.36	13049.21	7590414	7409369	1820.88	150	安　　徽
8770.14	1764.83	3191141	3171545	1013.77	38	福　　建
7277.95	2872.26	3967849	3914459	963.60	49	江　　西
70804.94	21622.47	17347442	17084509	3865.13	398	山　　东
37225.42	17052.33	11325028	11151714	2508.10	214	河　　南
53506.03	16319.18	9661879	9533522	2070.08	191	湖　　北
13486.41	8330.54	6021367	5940545	1513.96	68	湖　　南
20385.83	10054.21	13952573	13791629	4165.33	79	广　　东
7612.76	1215.55	3191035	3164522	811.13	42	广　　西
10726.32	561.66	1024895	1013355	275.02	30	海　　南
50313.10	12065.50	8186553	7891512	1556.84	117	重　　庆
112249.18	23888.19	15032000	14551915	2949.75	278	四　　川
6464.84	1599.15	2429413	2402341	607.03	33	贵　　州
5286.14	694.03	1290327	1182171	527.02	35	云　　南
	100.00	127500	80000	43.00		西　　藏
24446.28	12102.26	7600261	7526460	1349.73	194	陕　　西
23751.94	839.99	2672069	2506221	589.87	107	甘　　肃
11681.52	5324.55	671439	655589	184.04	35	青　　海
21301.69	1295.19	1399833	1385988	270.49	111	宁　　夏
119269.61	3255.98	4052527	3970034	804.36	345	新　　疆
18132.25	671.23	568955	544449	98.91	62	新疆生产建设兵团

1-8-4　城市液化石油气(2022年)

地区名称 Name of Regions	储气能力 (吨) Gas Storage Capacity (ton)	供气管道长度 (公里) Length of Gas Supply Pipeline (km)	供气总量(吨)		
			合　计 Total	销售气量 Quantity Sold	居民家庭 Households
全　国　National Total	**884087.08**	**2547.37**	**7584585.86**	**7560545.24**	**4444193.70**
北　京　Beijing	16860.40	206.00	156679.00	156291.00	120088.00
天　津　Tianjin	3038.40		102545.55	102224.25	45301.28
河　北　Hebei	11924.63	107.23	86307.00	85892.20	59803.22
山　西　Shanxi	6546.00		63300.52	62869.52	47120.28
内　蒙　古　Inner Mongolia	7367.40		56580.35	55946.58	38270.61
辽　宁　Liaoning	33966.16	199.12	497149.43	496743.97	98443.87
吉　林　Jilin	20221.60	29.81	103575.01	102909.85	45998.06
黑　龙　江　Heilongjiang	11304.30	109.93	156965.51	156322.10	77486.52
上　海　Shanghai	19998.40	263.23	220376.60	220023.59	125858.03
江　苏　Jiangsu	49860.91	119.98	560086.14	558688.98	322508.05
浙　江　Zhejiang	33941.04	251.14	762242.32	759706.56	537447.54
安　徽　Anhui	15532.15	189.60	159205.82	158438.51	85157.66
福　建　Fujian	15857.11	216.48	292750.76	291743.13	159167.74
江　西　Jiangxi	19727.10		173056.20	171511.98	137042.63
山　东　Shandong	31355.01	1.16	259539.95	257281.84	143353.41
河　南　Henan	13088.12	4.52	155250.32	153922.56	130148.03
湖　北　Hubei	34546.36	90.07	280347.58	279042.52	167928.12
湖　南　Hunan	20057.00		250661.11	250132.92	178509.37
广　东　Guangdong	275731.07	564.25	2180559.88	2177471.41	1315159.16
广　西　Guangxi	148374.25	2.09	296882.59	296172.93	201144.06
海　南　Hainan	3520.60		71746.33	71721.63	64340.78
重　庆　Chongqing	4222.11		50826.17	50588.07	26708.29
四　川　Sichuan	18815.05	170.83	210088.35	208373.74	86794.18
贵　州　Guizhou	14644.69		111284.68	111091.59	48624.78
云　南　Yunnan	23110.12	16.93	139859.58	139597.10	57775.31
西　藏　Tibet	1047.50	1.40	8445.21	8444.26	7734.96
陕　西　Shaanxi	13183.80		54850.62	54519.24	46663.10
甘　肃　Gansu	7244.80		47765.65	47520.53	24862.53
青　海　Qinghai	2498.30		8881.48	8842.12	5944.12
宁　夏　Ningxia	3322.00	0.12	18926.20	18864.30	8443.80
新　疆　Xinjiang	2680.70	3.08	44708.03	44554.49	27724.70
新疆生产 建设兵团　Xinjiang Production and Construction Corps	500.00	0.40	3141.92	3091.77	2641.51

Urban LPG Supply(2022)

Total Gas Supplied (ton)		用气户数	居民家庭	用气人口	液化石油气汽车加气站	地区名称
燃气汽车 Gas-Powered Automobiles	燃气损失量 Loss Amount	（户） Number of Household with Access to Gas (unit)	 Households	（万人） Population with Access to Gas (10,000 persons)	（座） Gas Stations for LPG- Fueled Motor Vehicles (unit)	 Name of Regions
160448.70	**24040.62**	**36904161**	**33633520**	**9332.83**	**278**	**全　国**
	388.00	2218695	2204327	439.90		北　京
	321.30	446262	415146	72.98		天　津
	414.80	682454	630721	158.92		河　北
	431.00	244860	214673	31.36	8	山　西
1666.50	633.77	628593	528185	199.17	10	内 蒙 古
50464.30	405.46	1143839	1025788	263.18	49	辽　宁
6163.67	665.16	751418	659186	210.40	14	吉　林
34941.97	643.41	847670	783446	252.43	38	黑 龙 江
2433.78	353.01	2065515	1960024	494.65	4	上　海
10256.29	1397.16	2336705	1968789	320.39	10	江　苏
11610.67	2535.76	4250451	3834335	971.95	12	浙　江
210.00	767.31	517742	463873	123.12	2	安　徽
	1007.63	1362940	1289391	491.89		福　建
	1544.22	1017530	947343	238.38		江　西
17742.10	2258.11	1103939	984354	269.75	61	山　东
5345.33	1327.76	834124	795679	300.49	12	河　南
1985.00	1305.06	1543596	1404923	351.85	3	湖　北
89.00	528.19	1257423	1162595	380.47	2	湖　南
8067.01	3088.47	8236125	7565260	2346.55	8	广　东
	709.66	1681448	1618539	480.55		广　西
	24.70	457995	451354	56.66		海　南
	238.10	158353	110090	41.50		重　庆
1652.00	1714.61	443750	358107	98.88	14	四　川
	193.09	794390	688775	242.99		贵　州
	262.48	745366	630494	246.36		云　南
692.00	0.95	73520	71437	28.23	6	西　藏
150.00	331.38	358617	258216	73.61	3	陕　西
5215.00	245.12	240844	213137	58.58	7	甘　肃
300.50	39.36	40225	38495	18.40	6	青　海
	61.90	130519	104328	21.14		宁　夏
1213.32	153.54	268002	234145	43.15	8	新　疆
250.26	50.15	21251	18365	4.95	1	新疆生产建设兵团

1-9-1　全国历年城市集中供热情况

年 份 Year	供 热 能 力 Heating Capacity		供 热 总 量 Total Heat Supplied	
	蒸 汽 （吨/小时） Steam (ton/hour)	热 水 （兆瓦） Hot Water (mega watts)	蒸 汽 （万吉焦） Steam (10,000 GJ)	热 水 （万吉焦） Hot Water (10,000 GJ)
1981	754	440	641	183
1982	883	718	627	241
1983	965	987	650	332
1984	1421	1222	996	454
1985	1406	1360	896	521
1986	9630	36103	3467	2704
1987	16258	27601	6669	3650
1988	18550	32746	5978	4848
1989	20177	25987	6782	4334
1990	20341	20128	7117	21658
1991	21495	29663	8195	21065
1992	25491	45386	9267	26670
1993	31079	48437	10633	29036
1994	34848	52466	10335	32056
1995	67601	117286	16414	75161
1996	62316	103960	17615	56307
1997	65207	69539	20604	62661
1998	66427	71720	17463	64684
1999	70146	80591	22169	69771
2000	74148	97417	23828	83321
2001	72242	126249	37655	100192
2002	83346	148579	57438	122728
2003	92590	171472	59136	128950
2004	98262	174442	69447	125194
2005	106723	197976	71493	139542
2006	95204	217699	67794	148011
2007	94009	224660	66374	158641
2008	94454	305695	69082	187467
2009	93193	286106	63137	200051
2010	105084	315717	66397	224716
2011	85273	338742	51777	229245
2012	86452	365278	51609	243818
2013	84362	403542	53242	266462
2014	84664	447068	55614	276546
2015	80699	472556	49703	302110
2016	78307	493254	41501	318044
2017	98328	647827	57985	310300
2018	92322	578244	57731	323665
2019	100943	550530	65067	327475
2020	103471	566181	65054	345004
2021	118784	593226	68164	357715
2022	125543	600194	67113	361226

注：1981年至1995年热水供热能力计量单位为兆瓦/小时；1981年至2000年蒸汽供热总量计量单位为万吨。

National Urban Centralized Heating in Past Years

管道长度(公里) Length of Pipelines (km)		集中供热 面　　积 （万平方米) Heated Area (10,000 m²)	年　份 Year
蒸　汽 Steam	热　水 Hot Water		
79	280	1167	1981
37	491	1451	1982
67	586	1841	1983
71	761	2445	1984
76	954	2742	1985
183	1335	9907	1986
163	1576	15282	1987
209	2193	13883	1988
401	2678	19386	1989
157	3100	21263	1990
656	3952	27651	1991
362	4230	32832	1992
532	5161	44164	1993
670	6399	50992	1994
909	8456	64645	1995
9577	24012	73433	1996
7054	25446	80755	1997
6933	27375	86540	1998
7733	30506	96775	1999
7963	35819	110766	2000
9183	43926	146329	2001
10139	48601	155567	2002
11939	58028	188956	2003
12775	64263	216266	2004
14772	71338	252056	2005
14012	79943	265853	2006
14116	88870	300591	2007
16045	104551	348948	2008
14317	110490	379574	2009
15122	124051	435668	2010
13381	133957	473784	2011
12690	147390	518368	2012
12259	165877	571677	2013
12476	174708	611246	2014
11692	192721	672205	2015
12180	201390	738663	2016
276288		830858	2017
371120		878050	2018
392917		925137	2019
425982		988209	2020
461493		1060316	2021
493417		1112500	2022

Note: Heating capacity through hot water from 1981 to 1995 is measured with the unit of megawatts/hour; Heating capacity through steam from 1981 to 2000 is measured with the unit of 10,000 tons.

1-9-2 城市集中供热(2022年)

地区名称 Name of Regions	蒸汽 Steam						供热能力 (兆瓦) Heating Capacity (mega watts)	热电厂 供热 Heating by Co-Generation
	供热能力 (吨/小时) Heating Capacity (ton/hour)	热电厂 供热 Heating by Co-Generation	锅炉房 供热 Heating by Boilers	供热总量 (万吉焦) Total Heat Supplied (10,000 GJ)	热电厂 供热 Heating by Co-Generation	锅炉房 供热 Heating by Boilers		
全 国 National Total	125543	114317	8606	67113	62158	4450	600194	310271
北 京 Beijing							51621	9425
天 津 Tianjin	1875	1785	90	873	844	29	32282	10881
河 北 Hebei	6427	5836	470	4306	4104	143	49765	30288
山 西 Shanxi	19325	18380	945	11332	10499	833	30190	22253
内 蒙 古 Inner Mongolia	2596	2596		1993	1993		51782	40308
辽 宁 Liaoning	19455	18484	971	11515	11091	424	74606	26193
吉 林 Jilin	1622	1542	80	904	887	17	48340	20421
黑 龙 江 Heilongjiang	10655	10066	589	5695	5375	320	54716	32255
江 苏 Jiangsu	7648	7284	364	2118	1948	170	15	
安 徽 Anhui	3200	3200		2350	2350		200	
山 东 Shandong	29940	25582	1859	12554	11336	836	70851	50888
河 南 Henan	6158	5637	521	3173	2937	187	26310	20233
湖 北 Hubei	1943	1943		1541	1541		0	
四 川 Sichuan								
贵 州 Guizhou							280	
云 南 Yunan							450	
西 藏 Tibet							46	46
陕 西 Shaanxi	8075	5618	2457	4099	2942	1141	28524	11661
甘 肃 Gansu	795	795		610	610		19688	8389
青 海 Qinghai							8157	6900
宁 夏 Ningxia	2004	2004		1002	1002		7643	7155
新 疆 Xinjiang	1085	1085		770	770		39264	12565
新疆生产 建设兵团 Xinjiang Production and Construction Corps	2740	2480	260	2278	1928	350	5465	411

Urban Central Heating(2022)

热水　Hot Water				管道长度 （公里） Length of Pipelines (km)	一级管网 First Class	二级管网 Second Class	供热面积 （万平方米） Heated Area (10,000 m²)	住　宅 Housing	公共建筑 Public Building	地区名称 Name of Regions
锅炉房 供　热 Heating By Boilers	供热总量 （万吉焦） Total Heat Supplied (10,000 GJ)	热电厂 供　热 Heating by Co- Generation	锅炉房 供　热 Heating by Boilers							
216857	361226	207305	121275	493417	126599	366818	1112500	846194	252081	全　　国
	20347	5686		68277	4769	63508	71301	48400	22901	北　　京
21401	16845	7355	9490	36244	9272	26971	58648	45478	13170	天　　津
10967	30015	17279	5979	45163	10521	34641	96034	77186	15431	河　　北
7314	19263	15216	3691	26340	9507	16832	83070	61168	20141	山　　西
11097	33155	25953	7202	27021	6946	20075	68794	46959	19751	内　蒙　古
45574	55688	21185	31988	68091	14733	53357	145320	108472	35756	辽　　宁
26391	28833	15401	12111	36535	8518	28017	71941	51122	20509	吉　　林
21669	43001	25511	16682	25576	6342	19234	90416	64630	25540	黑　龙　江
15	8		8	425	408	17	3545	3545		江　　苏
	9			831	821	10	2684	1589	1089	安　　徽
17009	42569	30979	9669	97017	30217	66800	192312	163162	27760	山　　东
3988	14385	11244	1937	16205	10083	6122	64403	56736	7255	河　　南
0	0		0	651	365	286	2008	1736	12	湖　　北
				76	20	56	19		19	四　　川
76	63		6	55	29	26	205	117	88	贵　　州
50	89		30	470	153	317	188	100	88	云　　南
	110	110		300	40	260	184	55	129	西　　藏
12602	13435	7077	5266	5146	3418	1728	52777	42707	8590	陕　　西
10679	12591	5515	6838	14212	3174	11037	30702	21864	8353	甘　　肃
1258	3482	2660	822	1143	353	791	10397	6222	4175	青　　海
88	6210	5921	47	6673	1959	4714	15612	11745	3732	宁　　夏
25941	19864	10005	9036	14912	4117	10795	45635	29511	15721	新　　疆
739	1262	210	471	2055	832	1223	6305	3688	1872	新疆生产 建设兵团

三、居民出行数据
Data by Residents Travel

1-10-1　全国历年城市轨道交通情况
National Urban Rail Transit System in Past Years

年 份 Year	建成轨道交通 的城市个数 （个） Number of Cities with Completed Rail Transport Lines (unit)	建成轨道交通 线路长度 （公里） The Length of Completed Rail Transport Lines (km)	正在建设轨道 交通的城市个数 （个） Number of Cities with Rail Transport Lines under Construction (unit)	正在建设轨道 交通线路长度 （公里） The Length of Rail Transport Lines under Construction (km)
1978	1	23		
1979	1	23		
1980	1	23		
1981	1	23		
1982	1	23		
1983	1	23		
1984	2	47		
1985	2	47		
1986	2	47		
1987	2	47		
1988	2	47		
1989	2	47		
1990	2	47		
1991	2	47		
1992	2	47		
1993	2	47		
1994	2	47		
1995	3	63		
1996	3	63		
1997	3	63		
1998	4	81		
1999	4	81		
2000	4	117		

注：2000年及以前年份，大连、鞍山、长春、哈尔滨的有轨电车没有统计在内；2017年至2021年，对建成轨道交通线路长度的历史数据进行了修订。

Note: For the year 2000 and before, the streetcar systems in Dalian, Anshan, Changchun and Harbin City were not included when collecting data on the number of cities with completed rail transport lines and the length of lines. The length of completed rail transport lines from 2017 to 2021 has been revised.

1-10-1　续表　continued

年　份 Year	建成轨道交通 的城市个数 （个） Number of Cities with Completed Rail Transport Lines (unit)	建成轨道交通 线路长度 （公里） The Length of Completed Rail Transport Lines (km)	正在建设轨道 交通的城市个数 （个） Number of Cities with Rail Transport Lines under Construction (unit)	正在建设轨道 交通线路长度 （公里） The Length of Rail Transport Lines under Construction (km)
2001	5	172		
2002	5	200		
2003	5	347		
2004	7	400		
2005	10	444		
2006	10	621		
2007	10	775		
2008	10	855		
2009	10	838.88	28	1991.36
2010	12	1428.87	28	1741.07
2011	12	1672.42	28	1891.29
2012	16	2005.53	29	2060.43
2013	16	2213.28	35	2760.38
2014	22	2714.79	36	3004.37
2015	24	3069.23	38	3994.15
2016	30	3586.34	39	4870.18
2017	32	4515.91	50	4913.56
2018	34	5062.70	50	5400.25
2019	41	5980.55	49	5594.08
2020	42	7641.09	45	5093.55
2021	50	8571.43	48	5172.30
2022	55	9575.01	44	4802.89

1-10-2 城市轨道交通(建成)(2022年)

地区名称 Name of Regions	线路长度(公里) Length of Lines (km)										
	合计 Total	地铁 Subway	轻轨 Light Rail	单轨 Monorail	有轨 Cable Car	磁浮 Maglev	快轨 Fast Track	APM APM	按敷设方式 by Ways of Laying		
									地面线 Surface Lines	地下线 Underground Lines	高架线 Elevated Lines
全 国 National Total	9575.01	8637.71	216.29	124.77	443.87	59.85	88.62	3.90	668.57	7093.88	1812.56
北 京 Beijing	864.13	831.73			21.00	11.40			74.00	627.53	162.60
天 津 Tianjin	293.86	233.75	52.25		7.86				16.82	216.98	60.06
河 北 Hebei	76.40	76.40								76.40	
山 西 Shanxi	23.65	23.65								23.65	
内 蒙 古 Inner Mongolia	49.03	49.03							0.34	45.84	2.85
辽 宁 Liaoning	428.62	329.81			98.81				149.86	192.25	86.51
吉 林 Jilin	108.71	43.00	65.71						19.44	52.78	36.49
黑 龙 江 Heilongjiang	79.72	79.72								79.72	
上 海 Shanghai	831.58	795.39			6.29	29.90			16.40	550.37	264.81
江 苏 Jiangsu	1055.41	938.68			81.63		35.10		127.06	689.75	238.60
浙 江 Zhejiang	797.24	721.66	8.26		13.80		53.52		19.04	627.25	150.95
安 徽 Anhui	217.20	170.95		46.25						168.29	48.91
福 建 Fujian	212.01	212.01							2.58	197.81	11.62
江 西 Jiangxi	128.31	128.31							0.20	122.62	5.49
山 东 Shandong	407.96	399.19			8.77				11.44	275.21	121.31
河 南 Henan	238.36	197.65	40.71						1.77	219.66	16.93
湖 北 Hubei	536.87	460.82			76.05				68.58	347.57	120.72
湖 南 Hunan	238.93	190.74	29.64			18.55			2.52	189.39	47.02
广 东 Guangdong	1242.94	1198.78			40.26			3.90	49.76	1066.41	126.77
广 西 Guangxi	128.44	128.44								128.44	
海 南 Hainan	8.37				8.37				8.37		
重 庆 Chongqing	459.89	346.25	19.72	78.52	15.40				24.66	258.66	176.57
四 川 Sichuan	554.70	515.40			39.30				41.59	452.55	60.56
贵 州 Guizhou	75.71	75.71							3.31	66.43	5.97
云 南 Yunnan	179.31	165.91			13.40				15.70	141.60	22.01
西 藏 Tibet											
陕 西 Shaanxi	271.20	271.20							2.20	223.19	45.81
甘 肃 Gansu	38.84	25.91			12.93				12.93	25.91	
青 海 Qinghai											
宁 夏 Ningxia											
新 疆 Xinjiang	27.62	27.62								27.62	
新疆生产建设兵团 Xinjiang Production and Construction Corps											

Urban Rail Transit System (Completed)(2022)

车站数(个) Number of Stations (unit)				换乘站数(个) Number of Transfer Stations (unit)	配置车辆数(辆) Number of Vehicles in Service (unit)								地区名称 Name of Regions	
合计 Total	地面站 Surface Lines	地下站 Underground Lines	高架站 Elevated Lines		合计 Total	地铁 Subway	轻轨 Light Rail	单轨 Monorail	有轨 Cable Car	磁浮 Maglev	快轨 Fast Track	APM APM		
6375	630	4931	814	1752	51712	49659	385	732	750	158	21	7	全	国
511	39	401	71	189	9128	8958			50	120			北	京
225	20	174	31	59	1530	1370	152		8				天	津
63		63		20	486	486							河	北
23		23		7	144	144							山	西
44	1	40	3	10	312	312							内	蒙古
309	136	144	29	63	945	834			111				辽	宁
102	21	44	37	22									吉	林
66		66		10	822	822							黑	龙江
508	12	385	111	188	7474	7416			44	14			上	海
670	83	503	84	127	4401	4307			91		3		江	苏
484	17	405	62	136	4447	4344	68		17		18		浙	江
205	25	141	39	44	1602	1362		240					安	徽
157		154	3	47	846	846							福	建
103		99	4	24	906	906							江	西
223	12	166	45	60	1487	1480			7				山	东
175	7	167	1	62	371	338	33						河	南
384	87	229	68	78	3052	2962			90				湖	北
152	1	144	7	59	1496	1440	32			24			湖	南
770	66	643	61	227	4043	3948			88			7	广	东
104		104		20	135	135							广	西
15	15				14				14				海	南
281	22	155	104	84	2023	1416	100	492	15				重	庆
370	36	313	21	101	4478	4298			180				四	川
57	2	50	5	22	498	498							贵	州
129	15	107	7	40	479	461			18				云	南
													西	藏
192	1	170	21	43	388	388							陕	西
32	12	20		5	43	26			17				甘	肃
													青	海
													宁	夏
21		21		5	162	162							新	疆
													新疆生产建设兵团	

1-10-3　城市轨道交通(在建)(2022年)

地区名称 Name of Regions	线路长度(公里) Length of Lines (km)										
	合计 Total	地铁 Subway	轻轨 Light Rail	单轨 Monorail	有轨 Cable Car	磁浮 Maglev	快轨 Fast Track	APM APM	按敷设方式 by Ways of Laying		
									地面线 Surface Lines	地下线 Under-ground Lines	高架线 Elevated Lines
全　国　National Total	**4802.89**	**4357.68**	**197.49**		**106.48**	**4.46**	**136.78**		**193.76**	**3945.38**	**663.75**
北　京　Beijing	262.17	262.17							25.10	201.27	35.80
天　津　Tianjin	210.40	210.40								161.48	48.92
河　北　Hebei	45.30	45.30								45.30	
山　西　Shanxi	28.74	28.74								28.74	
内 蒙 古　Inner Mongolia											
辽　宁　Liaoning	187.89	187.89							0.49	162.14	25.26
吉　林　Jilin	124.93	120.44	4.49							120.44	4.49
黑 龙 江　Heilongjiang	12.99	12.99								12.99	
上　海　Shanghai	161.85	161.85								155.26	6.59
江　苏　Jiangsu	562.83	562.83							4.48	497.68	60.67
浙　江　Zhejiang	452.87	293.30	159.57						12.23	262.39	178.25
安　徽　Anhui	160.61	160.61							0.31	128.79	31.51
福　建　Fujian	228.78	166.38					62.40		23.40	163.60	41.78
江　西　Jiangxi	31.75	31.75								28.30	3.45
山　东　Shandong	388.02	351.92			36.10				20.93	338.37	28.72
河　南　Henan	217.07	183.64	33.43						0.32	208.99	7.76
湖　北　Hubei	146.72	146.72								144.09	2.63
湖　南　Hunan	83.76	79.30				4.46			0.92	59.70	23.14
广　东　Guangdong	628.54	578.24			50.30				48.57	552.66	27.31
广　西　Guangxi	3.90	3.90								3.90	
海　南　Hainan											
重　庆　Chongqing	281.43	265.03					16.40		7.97	214.66	58.80
四　川　Sichuan	256.28	178.22			20.08		57.98		22.98	185.87	47.43
贵　州　Guizhou	73.35	73.35								64.31	9.04
云　南　Yunnan	20.34	20.34								20.34	
西　藏　Tibet											
陕　西　Shaanxi	162.20	162.20							26.06	113.94	22.20
甘　肃　Gansu	9.06	9.06								9.06	
青　海　Qinghai											
宁　夏　Ningxia											
新　疆　Xinjiang	61.11	61.11								61.11	
新疆生产建设兵团　Xinjiang Production and Construction Corps											

Urban Rail Transit System(Under Construction) (2022)

合计 Total	地面站 Surface Lines	地下站 Under-ground Lines	高架站 Elevated Lines	换乘站数(个) Number of Transfer Stations (unit)	合计 Total	地铁 Subway	轻轨 Light Rail	单轨 Monorail	有轨 Cable Car	磁浮 Maglev	快轨 Fast Track	APM	地区名称 Name of Regions
2769	71	2479	219	933	21497	20786	456		56	9	190		全　国
124	11	104	9	67	3180	3180							北　京
143	1	122	20	58	1110	1110							天　津
41		41		12									河　北
24		24		7	162	162							山　西
													内　蒙古
135		119	16	43	1189	1189							辽　宁
86	1	78	7	27									吉　林
12		12		2									黑　龙江
83		82	1	32	1110	1110							上　海
373	2	356	15	144	3684	3684							江　苏
181	1	139	41	41	1540	1116	424						浙　江
89		80	9	25	798	798							安　徽
109		98	11	40	640	610					30		福　建
19		17	2	3	336	336							江　西
270	8	247	15	71	2065	2043			22				山　东
132	1	131		55	336	304	32						河　南
79		75	4	41	1072	1072							湖　北
46		38	8	17	541	532				9			湖　南
326	15	300	11	55	274	262			12				广　东
3		3											广　西
													海　南
132	3	105	24	60	975	951					24		重　庆
157	26	117	14	74	1632	1474			22		136		四　川
42		40	2	11	384	384							贵　州
17		17		5	33	33							云　南
													西　藏
88	2	76	10	19									陕　西
9		9		5	10	10							甘　肃
													青　海
													宁　夏
49		49		19	426	426							新　疆
													新疆生产建设兵团

1-11-1 全国历年城市道路和桥梁情况
National Urban Road and Bridge in Past Years

年 份 Year	道路长度 （公里） Length of Roads (km)	道路面积 （万平方米） Surface Area of Roads (10,000 m²)	防洪堤长度 （公里） Length of Flood Control Dikes (km)	人均城市道路面积 （平方米） Urban Road Surface Area Per Capita (m²)
1978	26966	22539	3443	2.93
1979	28391	24069	3670	2.85
1980	29485	25255	4342	2.82
1981	30277	26022	4446	1.81
1982	31934	27976	5201	1.96
1983	33934	29962	5577	1.88
1984	36410	33019	6170	1.84
1985	38282	35872	5998	1.72
1986	71886	69856	9952	3.05
1987	78453	77885	10732	3.10
1988	88634	91355	12894	3.10
1989	96078	100591	14506	3.22
1990	94820	101721	15500	3.13
1991	88791	99135	13892	3.35
1992	96689	110526	16015	3.59
1993	104897	124866	16729	3.70
1994	111058	137602	16575	3.84
1995	130308	164886	18885	4.36
1996	132583	179871	18475	4.96
1997	138610	192165	18880	5.22
1998	145163	206136	19550	5.51
1999	152385	222158	19842	5.91
2000	159617	237849	20981	6.13

注：1.自2006年起，人均城市道路面积按城区人口和城区暂住人口合计为分母计算，括号内为与往年同口径数据。
　　2.自2013年开始，不再统计防洪堤长度。
Notes: 1.Since 2006, urban road surface per capita has been calculated based on denominator which combines both permanent and temporary residents in urban areas, and the data in brackets are the same index calculated by the method of past years.
2.Starting from 2013, the data on the length of flood prevention dyke has been unavailable.

1-11-1　续表　continued

年　份 Year	道路长度 （公里） Length of Roads (km)	道路面积 （万平方米） Surface Area of Roads (10,000 m²)	防洪堤长度 （公里） Length of Flood Control Dikes (km)	人均城市道路面积 （平方米） Urban Road Surface Area Per Capita (m²)
2001	176016	249431	23798	6.98
2002	191399	277179	25503	7.87
2003	208052	315645	29426	9.34
2004	222964	352955	29515	10.34
2005	247015	392166	41269	10.92
2006	241351	411449	38820	11.04(12.36)
2007	246172	423662	32274	11.43
2008	259740	452433	33147	12.21
2009	269141	481947	34698	12.79
2010	294443	521322	36153	13.21
2011	308897	562523	35051	13.75
2012	327081	607449	33926	14.39
2013	336304	644155		14.87
2014	352333	683028		15.34
2015	364978	717675		15.60
2016	382454	753819		15.80
2017	397830	788853		16.05
2018	432231	854268		16.70
2019	459304	909791		17.36
2020	492650	969803		18.04
2021	532476	1053655		18.84
2022	552163	1089330		19.28

1-11-2 城市道路和桥梁 （2022年）

地区名称 Name of Regions	道路长度 （公里） Length of Roads (km)	建成区 in Built District	道路面积 （万平方米） Surface Area of Roads (10,000 m²)	人行道 面积 Surface Area of Sidewalks	建成区 in Built District
全 国 **National Total**	**552162.72**	**487922.99**	**1089330.10**	**239642.94**	**981032.16**
北 京 Beijing	8681.35		15374.80	2806.12	
天 津 Tianjin	9669.05	8319.41	18686.54	4281.72	15662.53
河 北 Hebei	19753.14	19017.28	42488.02	10321.50	40059.48
山 西 Shanxi	10191.31	9602.48	23954.60	5521.11	22755.82
内 蒙 古 Inner Mongolia	11310.56	9668.45	22989.70	5973.34	21955.29
辽 宁 Liaoning	24900.49	21950.46	45380.53	12129.98	41483.70
吉 林 Jilin	11537.87	10265.96	20578.94	5153.80	18827.15
黑 龙 江 Heilongjiang	14387.74	13488.35	22896.92	4816.89	22241.14
上 海 Shanghai	5988.00	5988.00	12375.00	3163.00	12375.00
江 苏 Jiangsu	54234.23	45867.73	95086.03	15103.79	81070.17
浙 江 Zhejiang	32945.80	27851.06	63786.09	14067.49	54734.77
安 徽 Anhui	20371.90	19787.94	48037.38	10358.79	46384.06
福 建 Fujian	16811.87	15423.06	33167.14	6446.05	31162.81
江 西 Jiangxi	14796.84	14408.20	31551.56	6658.69	30266.63
山 东 Shandong	53808.04	46837.57	109956.89	21931.19	95793.31
河 南 Henan	19531.83	18654.42	48219.81	11053.71	44455.56
湖 北 Hubei	23970.40	23557.07	47876.31	11836.48	47410.93
湖 南 Hunan	18907.27	17278.29	39463.01	9410.95	36963.69
广 东 Guangdong	58665.63	49975.22	99181.81	19079.68	88629.45
广 西 Guangxi	15822.05	15393.55	32557.36	6134.56	29010.93
海 南 Hainan	5235.75	4901.71	8421.81	2599.19	7789.13
重 庆 Chongqing	12530.92	12109.33	26922.40	8111.76	26241.38
四 川 Sichuan	29333.04	27540.65	58436.62	14911.32	55273.45
贵 州 Guizhou	14083.75	9742.07	24325.33	5926.89	19722.80
云 南 Yunnan	9417.17	9151.89	18751.30	4367.26	18948.10
西 藏 Tibet	1125.72	766.30	2113.31	718.19	1702.13
陕 西 Shaanxi	11031.72	8856.91	26027.18	6404.96	22819.95
甘 肃 Gansu	7213.27	7013.30	15176.63	3659.46	14740.73
青 海 Qinghai	1670.59	1452.01	4184.24	1011.29	3870.47
宁 夏 Ningxia	3009.62	2854.76	8290.49	1477.76	7664.43
新 疆 Xinjiang	9394.74	8629.33	19273.71	3448.43	17943.95
新疆生产 建设兵团 Xinjiang Production and Construction Corps	1831.06	1570.23	3798.64	757.59	3073.22

Urban Roads and Bridges(2022)

桥梁数 （座） Number of Bridges (unit)	大桥及 特大桥 Great Bridge and Grand Bridge	立交桥 Inter-section	道路照明 灯 盏 数 （盏） Number of Road Lamps (unit)	安装路灯 道路长度 （公里） Length of The Road with Street Lamp (km)	地下综合 管廊长度 （公里） Length of The Utility Tunnel (km)	新建地下 综合管廊长度 （公里） Length of The New-bwit Utility Tunnel (km)	地区名称 Name of Regions
86260	11692	6334	33524880	419806.39	7093.95	1638.46	全　国
2401	568	461	317412	8436.36	220.61		北　京
1312	258	181	434559	8595.23	21.30	0.26	天　津
2491	401	307	1124609	13703.82	218.54	67.06	河　北
1451	289	209	556803	6281.87	53.87	9.29	山　西
546	98	67	658045	7916.66	67.76		内　蒙　古
2066	332	229	1387771	16798.62	118.44	1.15	辽　宁
1025	228	121	622707	8320.57	215.92	42.29	吉　林
1214	279	274	793825	10199.99	25.84		黑　龙　江
3098	3	50	709134	5988.00	106.07	1.73	上　海
14521	979	411	3923921	42751.12	292.23	22.27	江　苏
14247	770	214	1982612	24574.01	299.66	280.33	浙　江
2414	285	336	1262908	16671.50	151.53	32.30	安　徽
2657	479	118	1049793	10140.57	463.54	76.04	福　建
1276	220	100	1024366	11770.47	179.40	7.49	江　西
6134	473	242	2268577	40653.71	902.30	76.54	山　东
1973	189	222	1160372	15016.04	168.08	174.55	河　南
2564	527	230	1054034	21187.49	524.87	152.96	湖　北
1454	481	134	958097	14662.85	196.94	15.60	湖　南
9350	2041	921	3849583	50732.50	292.08	67.46	广　东
1402	279	184	817816	8430.34	200.31	60.79	广　西
249	22	11	179869	2341.20	73.89	71.94	海　南
2754	664	366	921464	11004.33	61.25	15.63	重　庆
4158	650	388	2133378	23818.12	850.77	153.83	四　川
1148	312	84	859165	5261.25	70.77	10.27	贵　州
1436	339	94	814994	8053.92	343.74	84.58	云　南
65	8	1	35132	546.69	10.62	10.62	西　藏
916	225	178	788354	8427.54	550.04	68.11	陕　西
715	141	93	461466	4914.74	55.12	15.74	甘　肃
239	3	4	152771	1097.32	124.02	9.60	青　海
226	23	8	276557	2541.44	47.50		宁　夏
698	115	95	820095	7809.81	168.01	97.38	新　疆
60	11	1	124691	1158.31	18.93	12.65	新疆生产 建设兵团

四、环境卫生数据
Data by Environmental Health

1-12-1　全国历年城市排水和污水处理情况
National Urban Drainage and Wastewater Treatment in Past Years

年 份 Year	排水管道长度 （公里） Length of Drainage Pipelines (km)	污水年排放量 （万立方米） Annual Quantity of Wastewater Discharged (10,000 m³)	污水处理厂 Wastewater Treatment Plant		污水年处理量 （万立方米） Annual Treatment Capacity (10,000 m³)	污水处理率 （%） Wastewater Treatment Rate (%)
			座数 （座） Number (unit)	处理能力 （万立方米／日） Treatment Capacity (10,000 m³/day)		
1978	19556	1494493	37	64		
1979	20432	1633266	36	66		
1980	21860	1950925	35	70		
1981	23183	1826460	39	85		
1982	24638	1852740	39	76		
1983	26448	2097290	39	90		
1984	28775	2253145	43	146		
1985	31556	2318480	51	154		
1986	42549	963965	64	177		
1987	47107	2490249	73	198		
1988	50678	2614897	69	197		
1989	54510	2611283	72	230		
1990	57787	2938980	80	277		
1991	61601	2997034	87	317	445355	14.86
1992	67672	3017731	100	366	521623	17.29
1993	75207	3113420	108	449	623163	20.02
1994	83647	3030082	139	540	518013	17.10
1995	110293	3502553	141	714	689686	19.69
1996	112812	3528472	309	1153	833446	23.62
1997	119739	3514011	307	1292	907928	25.84
1998	125943	3562912	398	1583	1053342	29.56
1999	134486	3556821	402	1767	1135532	31.93
2000	141758	3317957	427	2158	1135608	34.25

注：1978年至1995年污水处理厂座数及处理能力均为系统内数。

Note: Number of wastewater treatment plants and treatment capacity from 1978 to 1995 are limited to the statistical figure in the
　　building sector.

1-12-1　续表　continued

年 份 Year	排水管道长度 （公里） Length of Drainage Pipelines (km)	污水年 排放量 （万立方米） Annual Quantity of Wastewater Discharged (10,000 m³)	污水处理厂 Wastewater Treatment Plant		污水年处理量 （万立方米） Annual Treatment Capacity (10,000 m³)	污水处理率 (%) Wastewater Treatment Rate (%)
			座数 （座） Number (unit)	处理能力 （万立方米／日） Treatment Capacity (10,000 m³/day)		
2001	158128	3285850	452	3106	1196960	36.43
2002	173042	3375959	537	3578	1349377	39.97
2003	198645	3491616	612	4254	1479932	42.39
2004	218881	3564601	708	4912	1627966	45.67
2005	241056	3595162	792	5725	1867615	51.95
2006	261379	3625281	815	6366	2026224	55.67
2007	291933	3610118	883	7146	2269847	62.87
2008	315220	3648782	1018	8106	2560041	70.16
2009	343892	3712129	1214	9052	2793457	75.25
2010	369553	3786983	1444	10436	3117032	82.31
2011	414074	4037022	1588	11303	3376104	83.63
2012	439080	4167602	1670	11733	3437868	87.30
2013	464878	4274525	1736	12454	3818948	89.34
2014	511179	4453428	1807	13087	4016198	90.18
2015	539567	4666210	1944	14038	4288251	91.90
2016	576617	4803049	2039	14910	4487944	93.44
2017	630304	4923895	2209	15743	4654910	94.54
2018	683485	5211249	2321	16881	4976126	95.49
2019	743982	5546474	2471	17863	5369283	96.81
2020	802721	5713633	2618	19267	5572782	97.53
2021	872283	6250763	2827	20767	6118956	97.89
2022	913508	6389707	2894	21606	6268888	98.11

1-12-2 城市排水和污水处理(2022年)

地区名称 Name of Regions	污水排放量(万立方米) Annual Quantity of Wastewater Discharged (10,000 m³)	排水管道长度(公里) Length of Drainage Piplines (km)	污水管道 Sewers	雨水管道 Rainwater Drainage Pipeline	雨污合流管道 Combined Drainage Pipeline	建成区 in Built District	污水处理厂 座数(座) Number of Wastewater Treatment Plant (unit)	二级以上 Second Level or Above	处理能力(万立方米/日) Treatment Capacity (10,000 m³/day)	二级以上 Second Level or Above
全 国 National Total	6389707	913508	420646	406960	85902	792600	2894	2696	21606.1	20576.0
北 京 Beijing	212346	20137	9248	9398	1491	6617	78	78	712.1	712.1
天 津 Tianjin	114459	23910	11134	11575	1200	23472	46	46	344.5	344.5
河 北 Hebei	183612	23531	11443	12088		22182	97	94	715.2	697.2
山 西 Shanxi	109735	14644	6647	7454	544	13669	53	42	367.0	319.7
内 蒙 古 Inner Mongolia	64568	15457	8228	6624	605	14190	40	37	246.2	231.7
辽 宁 Liaoning	334822	25589	7207	9102	9281	20738	142	99	1108.3	884.6
吉 林 Jilin	139950	14137	5772	6722	1643	12119	52	50	459.2	457.5
黑 龙 江 Heilongjiang	133420	13120	3723	5752	3645	12728	73	73	433.3	433.3
上 海 Shanghai	219737	22289	9464	11577	1247	22289	42	42	896.8	896.8
江 苏 Jiangsu	525717	94107	47223	42822	4062	76135	213	209	1684.6	1650.6
浙 江 Zhejiang	395453	64373	33319	30046	1009	50645	116	114	1341.4	1338.5
安 徽 Anhui	227265	37867	16922	19234	1711	34951	98	98	813.6	813.6
福 建 Fujian	171902	23723	11327	11561	835	21542	63	60	552.2	518.7
江 西 Jiangxi	134537	23120	10254	9477	3389	22667	79	68	456.8	404.8
山 东 Shandong	369891	74568	33291	40121	1156	68581	233	233	1493.1	1493.1
河 南 Henan	260026	34497	15144	15998	3355	31557	127	119	1043.3	969.3
湖 北 Hubei	321542	39303	15130	18637	5536	38040	118	113	962.2	930.7
湖 南 Hunan	264396	26431	9943	11440	5047	25213	100	85	860.1	776.4
广 东 Guangdong	958060	140221	70053	51592	18576	107852	342	329	2936.0	2855.5
广 西 Guangxi	176310	22537	8121	9518	4897	22214	76	73	515.1	509.6
海 南 Hainan	45292	7611	3253	3340	1018	5671	31	27	137.7	121.1
重 庆 Chongqing	155510	25504	12368	11651	1484	24874	85	80	461.0	434.5
四 川 Sichuan	310905	49484	23222	23245	3016	46605	193	179	955.0	922.0
贵 州 Guizhou	81265	15035	7946	5878	1210	12865	118	118	408.1	408.1
云 南 Yunnan	124374	19461	9261	8778	1422	16098	72	66	371.1	356.0
西 藏 Tibet	9767	999	377	335	287	552	11	7	33.2	13.2
陕 西 Shaanxi	172906	14757	6794	6721	1243	12786	72	61	573.1	511.1
甘 肃 Gansu	46479	8980	4688	2985	1306	8561	31	30	213.0	183.0
青 海 Qinghai	18614	3778	2053	1485	241	3585	14	14	62.9	62.9
宁 夏 Ningxia	29147	2454	624	607	1223	2179	23	20	139.8	127.3
新 疆 Xinjiang	66255	10028	5393	1021	3615	9595	43	27	258.8	192.3
新疆生产建设兵团 Xinjiang Production and Construction Corps	11445	1858	1074	176	609	1830	13	5	51.8	6.6

Urban Drainage and Wastewater Treatment(2022)

Wastewater Treatment Plant				Other Wastewater Treatment Facilities 其他污水处理设施		Total Quantity of Wastewater Treated 污水处理总量 (万立方米)	Recycled Water 市政再生水			Name of Regions 地区名称
处理量(万立方米) Quantity of Wastewater Treated (10,000 m³)	二级以上 Second Level or Above	干污泥产生量(吨) Quantity of Dry Sludge Produced (ton)	干污泥处置量(吨) Quantity of Dry Sludge Treated (ton)	处理能力(万立方米/日) Treatment Capacity (10,000 m³/day)	处理量(万立方米) Quantity of Wastewater Treated (10,000 m³)	(10,000 m³)	生产能力(万立方米/日) Recycled Water Production Capacity (10,000 m³/day)	利用量(万立方米) Annual Quantity of Wastewater Recycled and Reused (10,000 m³)	管道长度(公里) Length of Piplines (km)	
6165942	5879829	13698583	13615717	998.9	102946	6268888	7938.5	1795475	16412	全　国
204244	204244	1749091	1748853	21.4	4014	208258	712.1	120540	2234	北　京
111424	111424	157983	157822	4.2	1256	112680	344.5	41679	2414	天　津
181923	177879	391926	390938			181923	547.3	91682	847	河　北
107083	93999	250277	246415	4.0	985	108068	261.0	25132	656	山　西
62992	58389	179204	177983			62992	204.7	30167	1540	内　蒙　古
325713	258937	498316	447036	14.6	2528	328242	317.3	71356	694	辽　宁
136930	136442	192624	191890			136930	118.1	29244	102	吉　林
125592	125592	214425	214375	41.5	3765	129357	48.6	17900	92	黑　龙　江
215438	215438	408357	408357			215438				上　海
489605	481156	1272505	1271632	197.0	22560	512165	632.3	146353	576	江　苏
382647	382073	980652	979329	17.7	5436	388083	220.4	47121	287	浙　江
218797	218797	350654	349591	37.4	4079	222876	352.3	70216	263	安　徽
160662	151821	252539	252476	51.8	8783	169445	223.3	43187	141	福　建
130378	114936	163080	163048	4.1	949	131327	3.0	301	2	江　西
363946	363946	829434	829323	5.7	515	364460	829.5	189665	1298	山　东
258802	237405	501793	497985	2.0	2	258804	561.5	117129	753	河　南
296252	285953	453076	452186	68.4	17474	313727	291.5	62356	54	湖　北
259377	232309	358405	358405	14.5	170	259546	121.1	38660	128	湖　南
939520	913327	1364798	1363939	35.2	4520	944041	1051.3	384749	120	广　东
163849	161716	319760	319716	359.8	10307	174156	108.1	32192	54	广　西
44790	42118	64833	64821	0.5	119	44909	21.8	3016	140	海　南
152498	144142	253389	250663	2.9	455	152953	20.9	1997	124	重　庆
286406	277298	543209	541394	85.9	12774	299179	301.2	78341	434	四　川
80053	80053	129860	129794	1.1	307	80360	36.3	4967	32	贵　州
121237	116813	170039	170036	16.7	1912	123149	48.4	36628	525	云　南
9432	3965	8534	8476			9432	0.5	23		西　藏
167964	151238	1243690	1235622			167964	168.2	44098	519	陕　西
45454	44690	126272	126272	8.0		45454	60.9	9299	453	甘　肃
17850	17850	26714	26714			17850	19.1	4490	127	青　海
28867	26036	55907	55907			28867	58.8	13077	342	宁　夏
64783	48092	169184	166684	4.5	5	64788	214.7	33026	1386	新　疆
11431	1752	18051	18036	0.1	32	11463	39.9	6887	74	新疆生产建设兵团

1-13-1　全国历年城市市容环境卫生情况

年 份 Year	生活垃圾　Domestic Garbage			
	清运量 （万吨） Quantity of Collected and Transported (10,000 ton)	无害化处理场(厂)座数 （座） Number of Harmless Treatment Plants/ Grounds (unit)	无害化处理能力 （吨／日） Harmless Treatment Capacity (ton/day)	无害化处理量 （万吨） Quantity of Harmlessly Treated (10,000 ton)
1979	2508	12	1937	
1980	3132	17	2107	215
1981	2606	30	3260	162
1982	3125	27	2847	190
1983	3452	28	3247	243
1984	3757	24	1578	188
1985	4477	14	2071	232
1986	5009	23	2396	70
1987	5398	23	2426	54
1988	5751	29	3254	75
1989	6292	37	4378	111
1990	6767	66	7010	212
1991	7636	169	29731	1239
1992	8262	371	71502	2829
1993	8791	499	124508	3945
1994	9952	609	130832	4782
1995	10671	932	183732	6014
1996	10825	574	155826	5568
1997	10982	635	180081	6292
1998	11302	655	201281	6783
1999	11415	696	237393	7232
2000	11819	660	210175	7255
2001	13470	741	224736	7840
2002	13650	651	215511	7404
2003	14857	575	219607	7545
2004	15509	559	238519	8089
2005	15577	471	256312	8051
2006	14841	419	258048	7873
2007	15215	458	279309	9438
2008	15438	509	315153	10307
2009	15734	567	356130	11220
2010	15805	628	387607	12318
2011	16395	677	409119	13090
2012	17081	701	446268	14490
2013	17239	765	492300	15394
2014	17860	818	533455	16394
2015	19142	890	576894	18013
2016	20362	940	621351	19674
2017	21521	1013	679889	21034
2018	22802	1091	766195	22565
2019	24206	1183	869875	24013
2020	23512	1287	963460	23452
2021	24869	1407	1057064	24839
2022	24445	1399	1109435	24419

注：1.1980年至1995年垃圾无害化处理厂，垃圾无害化处理量为垃圾加粪便。
　　2.自2006年起，生活垃圾填埋场的统计采用新的认定标准，生活垃圾无害化处理数据与往年不可比。

National Urban Environmental Sanitation in Past Years

粪便清运量 (万吨) Volume of Soil Collected and Transported (10,000 ton)	公共厕所 (座) Number of Latrine (unit)	市容环卫专用车辆设备总数 (辆) Number of Vehicles and Equipment Designated for Municipal Environmental Sanitation (unit)	每万人拥有公厕 (座) Number of Latrine per 10,000 Population (unit)	年份 Year
2156	54180	5316		1979
1643	61927	6792		1980
1547	54280	7917	3.77	1981
1689	56929	9570	3.99	1982
1641	62904	10836	3.95	1983
1538	64178	11633	3.57	1984
1748	68631	13103	3.28	1985
2710	82746	19832	3.61	1986
2422	88949	21418	3.54	1987
2353	92823	22793	3.14	1988
2603	96536	25076	3.09	1989
2385	96677	25658	2.97	1990
2764	99972	27854	3.38	1991
3002	95136	30026	3.09	1992
3168	97653	32835	2.89	1993
3395	96234	34398	2.69	1994
3066	113461	39218	3.00	1995
2931	109570	40256	3.02	1996
2845	108812	41538	2.95	1997
2915	107947	42975	2.89	1998
2844	107064	44238	2.85	1999
2829	106471	44846	2.74	2000
2990	107656	50467	3.01	2001
3160	110836	52752	3.15	2002
3475	107949	56068	3.18	2003
3576	109629	60238	3.21	2004
3805	114917	64205	3.20	2005
2131	107331	66020	2.88(3.22)	2006
2506	112604	71609	3.04	2007
2331	115306	76400	3.12	2008
2141	118525	83756	3.15	2009
1951	119327	90414	3.02	2010
1963	120459	100340	2.95	2011
1812	121941	112157	2.89	2012
1682	122541	126552	2.83	2013
1552	124410	141431	2.79	2014
1437	126344	165725	2.75	2015
1299	129818	193942	2.72	2016
	136084	228019	2.77	2017
	147466	252484	2.88	2018
	153426	281558	2.93	2019
	165186	306422	3.07	2020
	184063	327512	3.29	2021
	193654	341628	3.43	2022

Notes: 1.Quantity of garbage disposed harmlessly from 1980 to 1995 consists of quantity of garbage and soil.

2.Since 2006, treatment of domestic garbage through sanitary landfill has adopted new certification standard,so the datas of harmless treatmented garbage are not compared with the past years.

1-13-2 城市市容环境卫生(2022年)

地区名称 Name of Regions	道路清扫保洁面积(万平方米) Surface Area of Roads Cleaned and Maintained (10,000 m²)	机械化 Mechanization	生活垃圾							
			清运量(万吨) Collected and Transported (10,000 ton)	处理量(万吨) Volume of Treated (10,000 ton)	无害化处理厂(场)数(座) Number of Harmless Treatment Plants/ Grounds (unit)	卫生填埋 Sanitary Landfill	焚烧 Incineration	其他 other	无害化处理能力(吨/日) Harmless Treatment Capacity (ton/day)	卫生填埋 Sanitary Landfill
全 国 National Total	1081814	866860	24444.72	24438.83	1399	444	648	307	1109435	215167
北 京 Beijing	17812	12464	740.57	740.57	37	7	11	19	31461	4491
天 津 Tianjin	14111	12514	309.08	309.08	19		13	6	19550	
河 北 Hebei	40731	37019	748.99	748.99	47	11	30	6	37631	3761
山 西 Shanxi	26583	21403	466.67	466.67	28	14	12	2	19870	5170
内 蒙 古 Inner Mongolia	26264	22366	348.56	348.53	30	23	7		14728	7628
辽 宁 Liaoning	48882	29721	994.39	990.24	49	23	18	8	38324	13172
吉 林 Jilin	19902	16212	435.71	435.71	40	23	14	3	22210	9280
黑 龙 江 Heilongjiang	28126	22394	507.86	507.86	42	27	11	4	22447	8697
上 海 Shanghai	19385	18881	890.13	890.13	26	1	13	12	37012	5000
江 苏 Jiangsu	79266	74112	1958.71	1958.71	74	12	44	18	78539	6905
浙 江 Zhejiang	58712	49077	1553.53	1553.53	81	2	51	28	82166	1418
安 徽 Anhui	50253	45943	745.42	745.42	53	11	26	16	36779	6896
福 建 Fujian	25298	19760	874.53	874.53	37	4	24	9	31223	2250
江 西 Jiangxi	30285	28175	527.70	527.70	30	2	18	10	23289	1150
山 东 Shandong	80285	69787	1724.03	1724.03	107	24	61	22	75849	12239
河 南 Henan	56412	49168	1087.87	1087.74	48	16	30	2	46735	6575
湖 北 Hubei	50307	39339	1032.62	1032.62	67	21	34	12	41223	7982
湖 南 Hunan	36743	30501	860.79	860.79	48	23	16	9	38153	13532
广 东 Guangdong	120512	75442	3280.64	3280.56	174	32	74	68	178697	35666
广 西 Guangxi	30506	19279	601.43	601.43	43	19	18	6	30459	8159
海 南 Hainan	9659	6787	264.80	264.80	12	1	9	2	15000	500
重 庆 Chongqing	25218	18042	678.77	678.77	38	15	16	7	28956	5384
四 川 Sichuan	57972	43129	1259.20	1258.94	48	13	27	8	45580	7980
贵 州 Guizhou	20632	19803	415.85	415.39	46	14	22	10	24517	6427
云 南 Yunnan	20973	17260	538.12	538.12	35	16	17	2	16970	3588
西 藏 Tibet	4991	2171	62.17	62.05	8	7	1		2330	1630
陕 西 Shaanxi	25782	22621	654.88	654.87	41	24	10	7	32046	14146
甘 肃 Gansu	14882	12566	266.34	266.34	31	16	10	5	12342	3742
青 海 Qinghai	3902	2827	117.68	117.04	10	9		1	2152	2032
宁 夏 Ningxia	10325	8660	117.25	117.25	10	3	4	3	7270	1590
新 疆 Xinjiang	22557	16230	345.10	345.10	30	22	6	2	13940	6690
新疆生产建设兵团 Xinjiang Production and Construction Corps	4547	3207	35.32	35.32	10	9	1		1987	1487

Urban Environmental Sanitation(2022)

	Domestic Garbage					公共厕所 （座） Number of Latrines (unit)	三类 以上 Grade III and Above	市容环卫 专用车辆 设备总数 （辆） Number of Vehicles and Equipment Designated for Munici- pal Environ- mental Sanitation (unit)	地区名称 Name of Regions
焚烧 Inciner- ation	其他 other	无害化 处理量 （万吨） Volume of Harmlessly Treated (10,000 ton)	卫生 填埋 Sanitary Landfill	焚烧 Inciner- ation	其他 other				
804670	89598	24419.31	3043.20	19502.08	1874.02	193654	167975	341628	全　国
17150	9820	740.57	43.10	491.16	206.31	6435	6435	12159	北　京
18200	1350	309.08		287.35	21.73	4563	4253	5475	天　津
31850	2020	748.99	78.18	630.17	40.64	8252	7697	13600	河　北
14100	600	466.67	137.90	316.89	11.89	4430	3336	6931	山　西
7100		348.53	187.97	160.56		6956	5257	6856	内 蒙 古
23330	1822	990.24	265.11	672.38	52.74	5743	4150	13741	辽　宁
12250	680	435.71	86.20	342.97	6.54	4813	3681	9056	吉　林
12750	1000	507.86	195.26	301.85	10.75	5994	3442	11382	黑 龙 江
23000	9012	876.25	16.49	664.36	195.40	6253	3952	10054	上　海
64028	7606	1958.71	27.40	1734.85	196.46	14627	13766	25137	江　苏
72700	8048	1553.53		1366.41	187.12	10201	9220	13526	浙　江
27160	2723	745.42	4.19	683.15	58.08	7092	6937	12107	安　徽
26178	2795	874.53	33.02	785.99	55.52	6928	5590	9182	福　建
20850	1289	527.70	20.28	484.97	22.45	6049	6048	12168	江　西
59660	3950	1724.03	16.95	1629.63	77.45	10829	10151	22803	山　东
39410	750	1084.36	177.87	903.24	3.26	12684	12247	19947	河　南
30347	2894	1032.62	179.30	791.00	62.32	7889	6552	15201	湖　北
21631	2990	860.79	208.07	599.54	53.18	5168	3874	7933	湖　南
127935	15096	3279.03	289.79	2664.07	325.17	13515	12927	31110	广　东
19950	2350	601.43	95.24	478.63	27.56	3039	1863	13044	广　西
13400	1100	264.80		248.41	16.39	1475	1466	12467	海　南
19422	4150	678.77	60.42	527.68	90.67	4991	3936	4886	重　庆
36112	1488	1258.94	101.37	1112.89	44.68	9786	8377	15227	四　川
16750	1340	415.36	72.72	320.17	22.46	4324	3635	5974	贵　州
12707	675	537.41	97.75	427.12	12.54	6495	6311	6544	云　南
700		62.05	38.28	23.77		897	101	1564	西　藏
15950	1950	654.87	205.21	422.11	27.55	6689	6459	5877	陕　西
7650	950	266.34	69.94	184.23	12.17	3092	2688	6021	甘　肃
	120	117.04	107.99		9.05	834	716	1048	青　海
5000	680	117.25	18.13	80.38	18.74	905	842	2771	宁　夏
6900	350	345.10	181.95	157.96	5.20	2345	1949	6930	新　疆
500		35.32	27.14	8.19		361	117	907	新疆生产 建设兵团

五、绿色生态数据
Data by Green Ecology

1-14-1　全国历年城市园林绿化情况
National Urban Landscaping in Past Years

计量单位：公顷　　　　　　　　　　　　　　　　　　　　　　　　　　　　　　　　　　　Measurement Unit: Hectare

年 份 Year	建成区绿化 覆盖面积 Built District Green Coverage Area	建 成 区 绿地面积 Built District Area of Green Space	公园绿地 面　积 Area of Public Recreational Green Space	公园面积 Park Area	人均公园 绿地面积 （平方米） Public Recreational Green Space Per Capita (m²)	建成区绿 化覆盖率 (%) Green Coverage Rate of Built District (%)	建成区 绿地率 (%) Green Space Rate of Built District (%)
1981		110037	21637	14739	1.50		
1982		121433	23619	15769	1.65		
1983		135304	27188	18373	1.71		
1984		146625	29037	20455	1.62		
1985		159291	32766	21896	1.57		
1986		153235	42255	30740	1.84	16.9	
1987		161444	47752	32001	1.90	17.1	
1988		180144	52047	36260	1.76	17.0	
1989		196256	52604	38313	1.69	17.8	
1990	246829		57863	39084	1.78	19.2	
1991	282280		61233	41532	2.07	20.1	
1992	313284		65512	45741	2.13	21.0	
1993	354127		73052	48621	2.16	21.3	
1994	396595		82060	55468	2.29	22.1	
1995	461319		93985	72857	2.49	23.9	
1996	493915	385056	99945	68055	2.76	24.43	19.05
1997	530877	427766	107800	68933	2.93	25.53	20.57
1998	567837	466197	120326	73198	3.22	26.56	21.81
1999	593698	495696	131930	77137	3.51	27.58	23.03
2000	631767	531088	143146	82090	3.69	28.15	23.67

注：1.自2006年起，"公共绿地"统计为"公园绿地"。
　　2.自2006年起，"人均公共绿地面积"统计为以城区人口和城区暂住人口合计为分母计算的"人均公园绿地面积"，括
　　　号内数据约为与往年同口径数据。
Notes: 1.Since 2006,Public Green Space is changed to Public Recreational Green Space.
　　　　2.Since 2006,Public Recreational Green Space Per Capita has been calculated based on denominator which combines both
　　　　　permanent and temporary residents in urban areas, and the datas in brackets are the same index but calculated by the method
　　　　　of past years.

1-14-1 续表 continued

计量单位: 公顷

Measurement Unit: Hectare

年 份 Year	建成区绿化覆盖面积 Built District Green Coverage Area	建 成 区绿地面积 Built District Area of Green Space	公园绿地面 积 Area of Public Recreational Green Space	公园面积 Park Area	人均公园绿地面积（平方米） Public Recreational Green Space Per Capita (m²)	建成区绿化覆盖率(%) Green Coverage Rate of Built District (%)	建成区绿地率(%) Green Space Rate of Built District (%)
2001	681914	582952	163023	90621	4.56	28.38	24.26
2002	772749	670131	188826	100037	5.36	29.75	25.80
2003	881675	771730	219514	113462	6.49	31.15	27.26
2004	962517	842865	252286	133846	7.39	31.66	27.72
2005	1058381	927064	283263	157713	7.89	32.54	28.51
2006	1181762	1040823	309544	208056	8.3(9.3)	35.11	30.92
2007	1251573	1110330	332654	202244	8.98	35.29	31.30
2008	1356467	1208448	359468	218260	9.71	37.37	33.29
2009	1494486	1338133	401584	235825	10.66	38.22	34.17
2010	1612458	1443663	441276	258177	11.18	38.62	34.47
2011	1718924	1545985	482620	285751	11.80	39.22	35.27
2012	1812488	1635240	517815	306245	12.26	39.59	35.72
2013	1907490	1719361	547356	329842	12.64	39.86	35.93
2014	2017348	1819960	582392	367926	13.08	40.22	36.29
2015	2105136	1907862	614090	383805	13.35	40.12	36.36
2016	2204040	1992584	653555	416881	13.70	40.30	36.43
2017	2314378	2099120	688441	444622	14.01	40.91	37.11
2018	2419918	2197122	723740	494228	14.11	41.11	37.34
2019	2522931	2285207	756441	502360	14.36	41.51	37.63
2020	2637533	2398085	797912	538477	14.78	42.06	38.24
2021	2732400	2492509	835659	647962	14.87	42.42	38.70
2022	2820978	2579720	868508	672753	15.29	42.96	39.29

1-14-2 城市园林绿化(2022年)

地区名称 Name of Regions		绿化覆盖面积 (公顷) Green Coverage Area (hectare)	建成区 Built District	绿地面积 (公顷) Area of Green Space (hectare)
全　国	National Total	4021177	2820978	3586020
北　京	Beijing	99008	99008	93558
天　津	Tianjin	51496	48573	47713
河　北	Hebei	121254	99242	100563
山　西	Shanxi	65043	57004	58288
内蒙古	Inner Mongolia	77105	53383	71573
辽　宁	Liaoning	223966	115071	150462
吉　林	Jilin	106916	67433	99451
黑龙江	Heilongjiang	81491	68452	74527
上　海	Shanghai	178186	47317	172647
江　苏	Jiangsu	353661	216648	319725
浙　江	Zhejiang	215651	144373	189757
安　徽	Anhui	148248	113324	132363
福　建	Fujian	98677	82731	91088
江　西	Jiangxi	88242	83444	80560
山　东	Shandong	318591	249959	280650
河　南	Henan	154043	141908	135170
湖　北	Hubei	133241	123019	117985
湖　南	Hunan	99626	89084	99439
广　东	Guangdong	587124	292976	539604
广　西	Guangxi	93391	76383	81236
海　南	Hainan	21291	17771	19683
重　庆	Chongqing	85446	73114	76584
四　川	Sichuan	162542	148548	143459
贵　州	Guizhou	118582	50345	100487
云　南	Yunnan	69943	55562	63270
西　藏	Tibet	7250	6966	6858
陕　西	Shaanxi	88721	66204	80269
甘　肃	Gansu	36994	35041	32834
青　海	Qinghai	9761	9135	9160
宁　夏	Ningxia	27901	20473	26418
新　疆	Xinjiang	85273	58798	78620
新疆生产 建设兵团	Xinjiang Production and Construction Corps	12513	9690	12019

注：本表中北京市的各项绿化数据均为该市调查面积内数据。

Urban Landscaping(2022)

建成区 Built District	公园绿地 面　　积 （公顷） Area of Public Recreational Green Space (hectare)	公园个数 （个） Number of Parks (unit)	门票免费 Free Parks	公园面积 （公顷） Park Area (hectare)	地区名称 Name of Regions
2579720	868508	24841	24276	672753	全　　国
93558	36900	612	577	36397	北　京
44880	11580	170	166	3438	天　津
90698	30776	1032	1009	22491	河　北
51979	17559	441	430	14574	山　西
49587	18220	479	467	15238	内　蒙　古
108736	31357	746	724	22580	辽　宁
61494	17329	482	480	13029	吉　林
62504	19333	479	467	13125	黑　龙　江
45783	22976	473	463	4227	上　海
199839	59388	1424	1334	35678	江　苏
131062	44861	1902	1860	26013	浙　江
102878	33198	834	830	22931	安　徽
76374	23077	725	713	15523	福　建
77454	20701	902	895	16930	江　西
226947	75582	1496	1453	51148	山　东
125057	44627	681	658	22114	河　南
112987	37428	737	719	22383	湖　北
81497	25315	756	745	19448	湖　南
264659	118509	4969	4933	165322	广　东
66694	15570	474	459	16921	广　西
16531	4075	167	160	3337	海　南
67814	28516	594	584	16671	重　庆
131724	44715	946	922	28692	四　川
47988	14958	451	444	15587	贵　州
50900	15027	1372	1297	12701	云　南
6590	1557	167	167	1281	西　藏
59681	18900	451	450	12271	陕　西
32028	11183	225	221	7057	甘　肃
8558	2830	71	70	1854	青　海
19619	6762	149	148	3865	宁　夏
54280	12783	223	220	7488	新　疆
9342	2913	211	211	2443	新疆生产 建设兵团

Note: All the greening-related data for Beijing Municipality in the table are those for the areas surveyed in the city.

县城部分

Statistics for County Seats

一．综合数据
General Data

2-1-1　全国历年县城市政公用设施水平
Level of Service Facilities of National County Seat in Past Years

年 份 Year	供 水 普及率 (%) Water Coverage Rate (%)	燃 气 普及率 (%) Gas Coverage Rate (%)	每万人拥 有公共交 通车辆 (标台) Motor Vehicle for Public Fransport Per 10,000 Persons (standardunit)	人均道路 面 积 (平方米) Road Surface Area Per Capita (m²)	污 水 处理率 (%) Wastewater Treatment Rate (%)	园 林 绿 化			每 万 人 拥有公厕 (座) Number of Public Lavatories per 10,000 Persons (unit)
						人均公园 绿地面积 (平方米) Public Recreational Green Space Per Capita (m²)	建成区绿化 覆 盖 率 (%) Green Coverage Rate of Built District (%)	建成区 绿地率 (%) Green Space Rate of Built District (%)	
2000	84.83	54.41	2.64	11.20	7.55	5.71	10.86	6.51	2.21
2001	76.45	44.55	1.89	8.51	8.24	3.88	13.24	9.08	3.54
2002	80.53	49.69	2.51	9.37	11.02	4.32	14.12	9.78	3.53
2003	81.57	53.28	2.52	9.82	9.88	4.83	15.27	10.79	3.59
2004	82.26	56.87	2.77	10.30	11.23	5.29	16.42	11.65	3.54
2005	83.18	57.80	2.86	10.80	14.23	5.67	16.99	12.26	3.46
2006	76.43	52.45	2.59	10.30	13.63	4.98	18.70	14.01	2.91
2007	81.15	57.33	3.07	10.70	23.38	5.63	20.20	15.41	2.90
2008	81.59	59.11	3.04	11.21	31.58	6.12	21.52	16.90	2.90
2009	83.72	61.66		11.95	41.64	6.89	23.48	18.37	2.96
2010	85.14	64.89		12.68	60.12	7.70	24.89	19.92	2.94
2011	86.09	66.52		13.42	70.41	8.46	26.81	22.19	2.80
2012	86.94	68.50		14.09	75.24	8.99	27.74	23.32	2.09
2013	88.14	70.91		14.86	78.47	9.47	29.06	24.76	2.77
2014	88.89	73.24		15.39	82.12	9.91	29.80	25.88	2.76
2015	89.96	75.90		15.98	85.22	10.47	30.78	27.05	2.78
2016	90.50	78.19		16.41	87.38	11.05	32.53	28.74	2.82
2017	92.87	81.35		17.18	90.21	11.86	34.60	30.74	2.93
2018	93.80	83.85		17.73	91.16	12.21	35.17	31.21	3.13
2019	95.06	86.47		18.29	93.55	13.10	36.64	32.54	3.28
2020	96.66	89.07		18.92	95.05	13.44	37.58	33.55	3.51
2021	97.42	90.32		19.68	96.11	14.01	38.30	34.38	3.75
2022	97.86	91.38		20.31	96.94	14.50	39.35	35.65	3.95

注：1.自2006年起，人均和普及率指标按县城人口和县城暂住人口合计为分母计算，以公安部门的户籍统计和暂住人口统计为准。

　　2."人均公园绿地面积"指标2005年及以前年份为"人均公共绿地面积"。

　　3.从2009年起，县城公共交通内容不再统计。

Notes: 1.Since 2006,figure in terms of per capita and coverage rate have been calculated based on denominater which combines both permanent and temporary residents in county seat areas. And the population should come from statistics of police.

　　　2.Since 2005, Public Green Space Per Capita is changed to be Public Recreational Green Space Per Capita.

　　　3.Since 2009,statistics on county seat public transport have been removed.

2-1-2 全国县城市政公用设施水平(2022年)

地区名称 Name of Regions	人口密度 (人/平方公里) Population Density (person/square kilometer)	人均日生活用水量 (升) Daily Water Consumption Per Capita (liter)	供水普及率 (%) Water Coverage Rate (%)	公共供水普及率 Public Water Coverage Rate	燃气普及率 (%) Gas Coverage Rate (%)	建成区供水管道密度 (公里/平方公里) Density of Water Supply Pipelines in Built District (kilometer/square kilometer)	人均道路面积 (平方米) Road Surface Area Per Capita (m²)	建成区路网密度 (公里/平方公里) Road in Built District (kilometer/square kilometer)
全 国 National Total	**2171**	**137.18**	**97.86**	**96.34**	**91.38**	**12.30**	**20.31**	**7.18**
河 北 Hebei	2777	105.89	100.00	99.62	98.97	12.07	25.46	8.99
山 西 Shanxi	3582	96.50	94.66	89.24	81.25	12.74	17.00	7.75
内 蒙 古 Inner Mongolia	857	106.75	98.83	97.74	92.61	11.68	33.97	6.98
辽 宁 Liaoning	1427	130.40	96.83	95.53	88.30	12.18	16.12	4.65
吉 林 Jilin	2361	105.43	96.74	95.52	87.80	13.34	17.77	6.27
黑 龙 江 Heilongjiang	3014	99.20	94.79	94.69	60.41	10.61	14.53	7.11
江 苏 Jiangsu	2050	178.58	100.00	99.85	100.00	14.52	22.87	6.61
浙 江 Zhejiang	906	227.78	100.00	100.00	100.00	21.48	25.63	9.52
安 徽 Anhui	1852	153.91	97.67	94.77	96.72	13.82	25.15	6.55
福 建 Fujian	2557	186.67	99.60	99.42	98.60	16.70	19.86	8.82
江 西 Jiangxi	4389	164.58	97.95	97.77	97.13	18.15	25.27	7.91
山 东 Shandong	1369	112.49	98.79	94.50	98.07	7.75	22.83	6.16
河 南 Henan	2587	112.62	97.68	93.57	93.69	7.55	18.68	5.80
湖 北 Hubei	3232	156.96	97.95	97.47	97.96	14.44	19.50	6.98
湖 南 Hunan	3644	166.56	98.26	97.79	94.52	15.51	14.83	7.18
广 东 Guangdong	1544	166.70	99.21	99.21	97.99	13.90	14.92	6.49
广 西 Guangxi	2644	181.85	99.59	98.10	99.21	14.05	21.79	9.13
海 南 Hainan	2189	243.57	99.56	99.53	96.68	5.91	34.43	5.08
重 庆 Chongqing	2592	136.30	99.41	99.41	98.14	12.44	11.82	7.16
四 川 Sichuan	1478	133.81	96.52	96.00	90.63	11.83	15.10	6.56
贵 州 Guizhou	2556	123.57	97.20	96.86	82.72	11.79	21.25	8.89
云 南 Yunnan	3963	126.60	95.40	94.59	60.66	13.80	19.26	8.32
西 藏 Tibet	2645	165.99	84.36	77.43	58.25	8.85	17.16	4.96
陕 西 Shaanxi	3869	108.55	96.49	93.57	90.72	7.44	16.07	7.20
甘 肃 Gansu	4980	85.47	96.84	96.55	77.02	9.79	15.49	6.12
青 海 Qinghai	1942	92.54	96.43	96.43	63.95	9.89	23.24	7.54
宁 夏 Ningxia	3863	111.91	99.56	99.56	77.35	10.09	24.64	6.94
新 疆 Xinjiang	3143	149.10	99.54	99.33	97.53	10.28	23.98	6.72

Level of National County Seat Service Facilities(2022)

建成区道路面积率(%) Road Surface Area Rate of Built District (%)	建成区排水管道密度(公里/平方公里) Density of Sewers in Built District (kilometer/square kilometer)	污水处理率(%) Wastewater Treatment Rate (%)	污水处理厂集中处理率 Centralized Treatment Rate of Wastewater Treatment Plants	人均公园绿地面积(平方米) Public Recreational Green Space Per Capita (m²)	建成区绿化覆盖率(%) Green Coverage Rate of Built District (%)	建成区绿地率(%) Green Space Rate of Built District (%)	生活垃圾处理率(%) Domestic Garbage Treatment Rate (%)	生活垃圾无害化处理率 Domestic Garbage Harmless Treatment Rate	地区名称 Name of Regions
13.80	10.66	96.94	96.15	14.50	39.35	35.65	99.82	99.24	全　　国
18.14	10.23	98.53	98.53	14.28	42.53	38.55	100.00	100.00	河　　北
14.73	11.32	97.71	97.71	12.35	40.94	36.46	99.27	95.55	山　　西
15.13	8.53	97.94	97.94	22.41	37.34	34.81	99.99	99.99	内　蒙　古
7.89	6.08	101.05	101.03	13.10	24.00	20.04	99.59	99.59	辽　　宁
11.66	10.21	98.34	98.34	17.66	40.87	37.12	100.00	100.00	吉　　林
8.64	7.20	96.61	96.61	15.46	33.61	29.90	100.00	100.00	黑　龙　江
13.44	12.42	88.39	88.39	15.35	42.82	40.00	100.00	100.00	江　　苏
16.55	16.88	98.07	97.91	16.05	44.35	40.16	100.00	100.00	浙　　江
15.07	12.21	96.61	95.85	16.14	40.60	37.05	100.00	100.00	安　　徽
14.56	14.51	97.44	96.16	16.27	43.81	40.29	100.00	100.00	福　　建
15.31	14.06	95.18	93.84	18.09	43.13	39.20	100.00	100.00	江　　西
13.29	10.44	98.15	98.15	16.80	41.66	37.82	100.00	100.00	山　　东
13.44	9.43	98.47	98.38	12.28	38.50	33.50	99.75	97.94	河　　南
14.32	10.24	96.28	96.28	14.09	40.96	37.58	100.00	100.00	湖　　北
12.74	10.62	96.47	96.33	11.31	37.85	34.65	99.98	99.98	湖　　南
10.51	7.09	95.60	95.60	14.13	37.24	34.12	100.00	100.00	广　　东
16.10	13.08	98.25	92.68	13.09	38.64	34.11	100.00	100.00	广　　西
10.62	6.08	118.99	101.15	7.48	37.11	32.27	100.00	100.00	海　　南
13.19	16.42	100.44	100.44	15.19	43.90	40.19	100.00	100.00	重　　庆
13.14	10.54	95.64	91.22	14.63	39.69	36.02	99.83	99.79	四　　川
14.47	8.18	97.22	97.22	15.29	39.73	37.75	99.07	99.07	贵　　州
15.80	15.72	98.10	98.09	12.92	42.07	38.29	100.00	94.16	云　　南
5.73	7.13	64.30	62.94	1.43	6.41	4.05	99.14	95.43	西　　藏
12.40	8.77	95.45	95.45	11.14	36.56	32.41	99.42	99.42	陕　　西
11.29	10.12	97.16	97.16	13.67	32.31	28.60	99.96	99.96	甘　　肃
12.62	9.69	92.49	92.49	8.44	27.90	24.15	95.50	95.50	青　　海
14.69	9.00	98.86	98.86	18.45	40.74	37.60	100.00	100.00	宁　　夏
11.56	7.26	96.27	96.26	17.01	41.49	37.95	99.93	99.93	新　　疆

2-2-1　全国历年县城数量及人口、面积情况

面积计量单位：平方公里
人口计量单位：万人

年 份 Year	县个数 Number of Counties	县城人口 County Seat Population	县城暂住人口 County Seat Temporary Population
2000	1674	14157	
2001	1660	9012	
2002	1649	8874	
2003	1642	9235	
2004	1636	9641	
2005	1636	10030	
2006	1635	10963	934
2007	1635	11581	1011
2008	1635	11947	1079
2009	1636	12259	1120
2010	1633	12637	1236
2011	1627	12946	1393
2012	1624	13406	1514
2013	1613	13701	1566
2014	1596	14038	1615
2015	1568	14017	1598
2016	1537	13858	1583
2017	1526	13923	1701
2018	1519	13973	1722
2019	1516	14111	1755
2020	1495	14055	1791
2021	1482	13941	1714
2022	1481	13836	1773

National Changes in Number of Counties, Population and Area in Past Years

Area Measurement Unit: Square Kilometer

Population Measurement Unit: 10,000 persons

县城面积 County Seat Area	建成区面积 Area of Built District	城市建设 用地面积 Area of Urban Construction Land	年 份 Year
53197	13135	8788	2000
57651	10427	9157	2001
56138	10496	9455	2002
53197	11115	10180	2003
53649	11774	11106	2004
63382	12383	12383	2005
76508	13229	13456	2006
93887	14260	14680	2007
130813	14776	15534	2008
154603	15558	15671	2009
175926	16585	16405	2010
93567	17376	16151	2011
94834	18740	17437	2012
86225	19503	17935	2013
79946	20111	18694	2014
75204	20043	18718	2015
72591	19467	18242	2016
71583	19854	18864	2017
70357	20238	19071	2018
76044	20672	19427	2019
75197	20867	19632	2020
72468	21026	19752	2021
71912	21092	19839	2022

2-2-2 全国县城人口和建设用地(2022年)

面积计量单位: 平方公里
人口计量单位: 万人

地区名称 Name of Regions	县面积 County Area	县人口 County Permanent Population	县暂住 人口 County Temporary Population	县城面积 County Seat Area	县城人口 County Seat Permanent Population	县城暂 住人口 County Seat Temparory Population	建成区 面积 Area of Built District	小 计 Subtotal
全 国 National Total	7264494	61970	3136	71911.72	13836.12	1772.86	21091.56	19839.08
河 北 Hebei	136224	3980	158	3774.11	946.07	101.85	1423.78	1371.85
山 西 Shanxi	124002	1834	89	1778.25	577.54	59.51	701.95	666.99
内 蒙 古 Inner Mongolia	1049033	1424	126	5563.75	406.16	70.66	1017.41	934.60
辽 宁 Liaoning	74788	999	26	1435.88	193.52	11.43	382.06	353.41
吉 林 Jilin	84796	697	25	680.91	146.79	14.00	236.38	235.98
黑 龙 江 Heilongjiang	217562	1303	36	1167.52	323.98	27.89	556.51	521.73
江 苏 Jiangsu	33460	1946	63	2534.03	486.69	32.86	737.38	718.90
浙 江 Zhejiang	49772	1410	328	5055.42	345.02	113.02	636.48	649.00
安 徽 Anhui	92536	4259	173	4960.25	807.99	110.52	1348.07	1322.49
福 建 Fujian	75600	1710	181	1827.00	400.06	67.08	562.95	532.47
江 西 Jiangxi	121503	2757	82	1664.98	679.45	51.31	1088.13	1055.06
山 东 Shandong	65130	3324		7568.30	981.96	54.37	1587.90	1519.78
河 南 Henan	116682	6946	210	5849.30	1390.15	122.87	1905.85	1788.67
湖 北 Hubei	92369	1917	97	1442.55	418.23	48.05	598.62	550.98
湖 南 Hunan	155622	4276	307	3194.30	963.72	200.20	1267.93	1204.21
广 东 Guangdong	79398	2210	114	3345.88	469.79	46.96	650.91	579.34
广 西 Guangxi	158193	3071	77	1959.70	468.37	49.72	676.97	641.61
海 南 Hainan	17400	323	24	279.20	51.68	9.43	167.25	139.45
重 庆 Chongqing	39137	925	113	823.39	182.69	30.75	190.92	172.88
四 川 Sichuan	399983	4714	278	8171.85	1022.33	185.77	1312.81	1213.98
贵 州 Guizhou	135187	2872	108	2622.63	601.89	68.43	879.93	757.58
云 南 Yunnan	296315	3072	147	1560.85	541.07	77.56	725.13	694.34
西 藏 Tibet	1194990	269	43	323.15	59.87	25.60	167.73	133.83
陕 西 Shaanxi	145453	1945	80	1356.77	468.85	56.05	634.13	592.27
甘 肃 Gansu	364950	1807	82	835.53	377.66	38.47	508.81	473.65
青 海 Qinghai	481607	352	28	588.35	98.77	15.48	202.47	172.90
宁 夏 Ningxia	37579	318	27	310.24	104.63	15.21	191.80	186.83
新 疆 Xinjiang	1425225	1306	113	1237.63	321.19	67.81	731.30	654.30

County Seat Population and Construction Land(2022)

Area Measurement Unit: Square Kilometer
Population Measurement Unit: 10,000persons

城市建设用地面积 Area of Urban Construction Land								本年征用土地面积	耕 地	地区名称
居 住 用 地 Residential	公共管理与公共服务用地 Administration and Public Services	商业服务业设施用地 Commercial and Business Facilities	工业用地 Industrial, Manufacturing	物流仓储用地 Logistics and Warehouse	道路交通设施用地 Road, Street and Transportation	公用设施用地 Municipal Utilities	绿 地 与 广场用地 Green Space and Square	Area of Land Requisition This Year	Arable Land	Name of Regions
6626.10	1710.64	1252.27	2547.29	451.70	3210.26	673.46	3367.36	871.90	405.26	全　　国
436.29	90.86	81.16	92.97	18.17	275.34	24.00	353.06	45.67	26.91	河　　北
260.64	56.12	35.75	45.53	12.18	110.94	19.66	126.17	5.27	2.52	山　　西
292.60	92.17	67.26	83.14	19.20	165.82	33.01	181.40	13.13	2.56	内 蒙 古
146.12	22.96	19.76	73.43	8.61	42.40	13.89	26.24	7.36	3.30	辽　　宁
86.44	16.58	13.56	29.37	9.23	32.67	7.91	40.22	8.06	3.78	吉　　林
216.87	39.79	27.26	79.04	24.09	66.64	14.97	53.07	21.29	7.60	黑 龙 江
237.74	53.06	41.04	133.04	7.80	125.16	18.83	102.23	51.31	33.13	江　　苏
201.05	56.39	43.21	128.99	8.11	97.33	19.12	94.80	41.10	19.57	浙　　江
381.82	89.67	81.09	222.50	34.18	248.28	43.68	221.27	111.73	55.48	安　　徽
193.24	46.17	30.22	85.15	5.90	87.53	14.16	70.10	24.88	6.67	福　　建
319.58	94.79	66.63	164.81	21.90	179.60	30.30	177.45	58.92	17.45	江　　西
516.14	123.89	106.32	306.15	35.62	204.76	38.98	187.92	49.70	25.15	山　　东
564.97	139.39	99.41	197.05	43.27	321.68	66.86	356.04	45.53	26.13	河　　南
179.89	54.03	34.66	83.44	11.25	86.72	19.98	81.01	37.03	17.74	湖　　北
428.36	113.11	86.73	157.93	40.64	164.25	47.77	165.42	42.91	10.57	湖　　南
214.12	49.88	37.25	90.15	9.69	76.96	23.40	77.89	15.88	3.50	广　　东
199.29	58.18	31.14	75.67	14.71	117.32	16.89	128.41	49.32	17.07	广　　西
42.42	12.56	9.36	27.37	4.07	25.92	2.38	15.37	4.61	1.19	海　　南
63.88	16.66	9.29	19.16	2.49	28.87	6.13	26.40	12.82	7.26	重　　庆
403.72	117.08	73.16	178.08	33.95	169.38	53.18	185.43	70.74	36.94	四　　川
285.44	76.60	56.39	79.60	21.24	107.67	33.42	97.22	24.31	9.24	贵　　州
213.55	68.66	47.86	43.25	14.09	125.89	27.30	153.74	71.42	41.68	云　　南
44.81	19.93	12.39	7.08	5.30	22.23	11.14	10.95	4.22	2.27	西　　藏
186.00	48.17	37.77	37.36	13.17	98.77	26.97	144.06	13.57	3.84	陕　　西
176.82	51.28	34.19	29.07	12.09	71.56	22.05	76.59	24.51	14.18	甘　　肃
57.23	19.24	11.43	16.91	3.65	29.38	9.04	26.02	1.66	1.11	青　　海
65.14	17.42	16.38	14.97	5.51	31.19	6.57	29.65	8.61	5.65	宁　　夏
211.93	66.00	41.60	46.08	11.59	96.00	21.87	159.23	6.34	2.77	新　　疆

2-3-1　全国历年县城维护建设资金收入

计量单位：万元

年　份 Year	合　计 Total	城市维护 建设税 Urban Maintenance and Construction Tax	城市公用 事业附加 Extra- Charges for Municipal Utilities	中央财政 拨　款 Financial Allocation From Central Government Budget	地方财政 拨　款 Financial Allocation From Local Government Budget
2000	1475288	218874	45849	83404	209840
2001	2339413	284380	47428	68460	378484
2002	3165232	319198	49195	163717	552206
2003	4703331	418992	52225	249529	829818
2004	5178923	482254	58156	159856	1080049
2005	6043627	562226	66971	173570	1405247
2006	5280388	624690	87843	281139	121727
2007	7218834	827944	115565	255958	2022760
2008	9736493	1109612	217957	647527	2927932
2009	13820241	1272267	165937	1125862	5319207
2010	20582038	1520894	215283	1410514	5465372
2011	26240827	1989632	393902	1878689	7259456
2012	31833379	2529027	403033	2937679	9382874
2013	36096026	2840257	523755	2811897	
2014	42036116	2984908	488213	2469729	
2015	36400093	2818465	479671	2199425	
2016	36538509	2940675	473941	2177512	

注：自2006年起，县城维护建设资金收入仅包含财政性资金，不含社会融资。地方财政拨款中包括省、市财政专项拨款和市级以下财政资金；其他收入中包括市政公用设施配套费、市政公用设施有偿使用费、土地出让转让金、资产置换收入及其他财政性资金。

National Revenue of County Seat Maintenance and Construction Fund in Past Years

Measurement Unit: 10,000 RMB

水资源费 Water Resource Fee	国内贷款 Domestic Loan	利用外资 Foreign Investment	企事业单位自筹资金 Self-Raised Funds by Enterprises and Institutions	其他收入 Other Revenues	年 份 Year
5421	177939	52215	243841	437904	2000
7807	261940	65162	535359	690393	2001
9087	374804	76792	775543	844690	2002
14283	603166	131672	1261785	1141861	2003
13112	645288	134733	1203808	1401667	2004
16618	698906	158159	1440598	1521332	2005
30201				4134788	2006
55081				3941526	2007
59435				4774030	2008
88990				5847978	2009
116924				11853051	2010
267072				14452076	2011
283277				16297489	2012
					2013
					2014
					2015
					2016

Note: Since 2006, national revenue of county seats'maintenance and construction fund includes the fund fiscal budget, not including social funds. Local Financial Allocation include province and city special financial allocation, and Other Revenues include fees for expansion of municipal utilites capacity, fees for use of municipal utilities, land transfer income, asset replacement income and other financial funds.

2-3-2 全国历年县城维护建设资金支出

计量单位: 万元

| 年 份 Year | 支出合计 Total | 按用途分 By Purpose | | | 供 水 Water Supply | 燃 气 Gas Supply | 集中供热 Central Heating |
		固定资产投资支出 Expenditure from Investment in Fixed Assets	维护支出 Maintenance Expenditure	其他支出 Other Expenditures			
2000	1543701	1224032	280838	38831	159991	17394	44294
2001	2336929	1746365	333699	256865	203748	42053	81911
2002	3196841	2529827	389298	277716	237120	82809	100698
2003	4661734	3907547	459527	294660	350503	120290	176962
2004	5077874	4221092	559888	296894	418492	132216	226354
2005	5927238	4831130	659689	436419	486596	187305	270791
2006	4971017	3660147	858483	452387	335755	84706	138289
2007	6806308	4796581	1207586	802141	356545	91517	227936
2008	9192126	6722788	1434309	1020577	429065	161977	322488
2009	13182941	10266437	1773499	1322453	581938	195643	471494
2010	20451108	16006075	2348570	2096535	764970	347357	607139
2011	24246084	18101051	3240910	2904123			
2012	33299318	23820534	3899428	5579356			
2013	29418982						
2014	29798372						
2015	27900620						
2016	28732749						

注: 从2009年起, 县城公共交通内容不再统计。

National Expenditure of County Seat Maintenance and Construction Fund in Past Years

Measurement Unit: 10,000 RMB

按行业分　By Indurstry							年　份
公共交通 Public Transportation	道路桥梁 Road and Bridge	排　水 Sewerage	防　洪 Flood Control	园林绿化 Landscaping	市容环境卫生 Environmental Sanitation	其　他 Other	Year
21325		113500		81017	69388	226679	2000
56570	900589	170371	84847	167091	105395	524354	2001
69067	1267934	256982	107759	213189	129337	731946	2002
92844	2040028	401354	149805	293000	208242	828706	2003
110896	2068361	447482	164216	386914	206046	916897	2004
136966	2390132	545025	192193	419119	239892	1059219	2005
93385	2192650	533822	144505	434397	284086	729422	2006
164390	2897997	781286	191378	605518	398862	1090879	2007
232359	3904437	1098048	183828	964899	524292	1370733	2008
	5120827	2278356	299657	1375296	1009819	1849911	2009
	8356614	2383811	418408	2543138	1380225	7444422	2010
							2011
							2012
							2013
							2014
							2015
							2016

Note: Since 2009, statistics on county seat public transport have been removed.

2-4-1　按行业分全国历年县城市政公用设施建设固定资产投资

计量单位: 亿元

年 份 Year	本年固定资产投资总额 Completed Investment of This Year	供 水 Water Supply	燃 气 Gas Supply	集中供热 Central Heating	公共交通 Public Transport	轨道交通 Rail Transit System	道路桥梁 Road and Bridge
2001	337.4	23.6	6.2	8.3	10.0		117.2
2002	412.2	28.1	10.5	13.2	10.6		152.9
2003	555.7	38.8	13.9	18.5	11.8		228.3
2004	656.8	44.4	15.1	24.3	12.3		246.7
2005	719.1	53.4	21.9	29.8	14.1		285.1
2006	730.5	44.3	24.1	28.9	10.3		319.1
2007	812.0	42.3	26.9	42.4	17.7		358.1
2008	1146.1	48.2	35.7	58.5	18.3		532.1
2009	1681.4	78.2	37.0	72.9			690.3
2010	2569.8	88.1	67.2	124.2			1132.1
2011	2859.6	127.6	112.7	155.7			1393.4
2012	3984.7	146.6	137.3	167.8			1934.7
2013	3833.6	164.9	182.3	223.4			1923.5
2014	3572.9	172.6	158.1	187.6			1908.4
2015	3099.8	156.4	112.6	171.0			1663.9
2016	3394.5	160.7	123.1	180.7			1805.8
2017	3634.2	226.3	121.0	194.1			1603.3
2018	3026.0	144.1	103.5	158.6		16.3	1185.0
2019	3076.7	168.1	136.2	133.8		11.1	1312.2
2020	3884.3	232.2	79.7	129.8		14.6	1399.6
2021	4087.2	255.2	75.6	161.0		9.7	1470.6
2022	4290.8	289.4	84.5	177.3		1.9	1518.6

注: 1.自2009年开始, 全国县城市政公用设施建设固定资产投资中不再包括公共交通固定资产投资。
　　2.自2013年开始, 全国县城市政公用设施建设固定资产投资中不再包括防洪固定资产投资。

National Fixed Assets Investment of County Seat Service Facilities by Industry in Past Years

Measurement Unit: 100 million RMB

排 水 Sewerage	污水处理 及 其 再 生 利 用 Wastewater Treatment and reused	防 洪 Flood Control	园林绿化 Land- scaping	市容环境 卫　生 Environmental Sanitation	垃圾处理 Garbage Treatment	地下综合 管　廊 Utility Tunnel	其 他 Other	年 份 Year
20.4	5.3	11.8	18.2	6.9	1.6		114.9	2001
33.0	12.9	13.7	22.0	10.6	3.9		117.7	2002
44.6	16.1	17.5	30.5	14.9	7.0		136.6	2003
52.5	17.2	19.1	41.0	14.7	4.5		186.7	2004
63.5	24.6	20.4	45.0	17.0	5.4		169.8	2005
72.1	37.0	11.6	46.2	42.2	7.6		132.4	2006
107.1	67.2	17.5	76.0	29.2	16.4		94.9	2007
141.2	80.8	26.4	174.1	37.2	23.2		74.4	2008
305.7	225.7	42.1	222.8	94.7	62.9		137.6	2009
271.1	165.7	45.8	373.6	121.9	83.4		345.9	2010
201.6	108.8	49.3	445.7	172.2	67.2		201.5	2011
229.6	105.0	69.8	581.4	274.6	136.0		442.8	2012
276.1	114.9		587.4	97.3	44.2		378.7	2013
296.1	139.2		521.0	97.4	36.0		231.9	2014
265.8	113.3		480.8	74.0	31.6		175.3	2015
263.0	114.6		500.7	115.9	52.4	22.9	221.8	2016
383.9	104.7		630.6	114.9	53.5	73.5	286.6	2017
367.7	168.0		558.7	134.6	72.2	45.2	312.3	2018
366.6	176.0		482.5	127.3	83.3	46.9	292.1	2019
560.9	306.2		568.2	267.4	199.5	36.9	594.9	2020
636.0	325.9		364.5	269.8	206.9	30.0	814.7	2021
771.7	319.3		352.5	222.5	160.6	33.3	839.0	2022

Notes: 1. Starting from 2009, the national fixed assets investment in the construction of municipal public utilities facilities no longer include the fixed assets investment in public transport.

2. Starting from 2013, the national fixed assets investment in the construction of municipal public utilities facilities no longer include the fixed assets investment in flood prevention.

2-4-2 按行业分全国县城市政公用设施建设固定资产投资(2022年)

计量单位: 万元

地区名称 Name of Regions	本年完成投资 Completed Investment of This Year	供 水 Water Supply	燃 气 Gas Supply	集中供热 Central Heating	轨道交通 Rail Transit System	道路桥梁 Road and Bridge	地下综合管廊 Utility Tunnel
全 国 National Total	42907659	2893654	845043	1772679	19313	15186401	332893
河 北 Hebei	3200883	112348	25274	206286	5	921587	36892
山 西 Shanxi	1395113	39197	28729	235240		504535	10839
内 蒙 古 Inner Mongolia	648104	65751	7890	113586	2547	128154	
辽 宁 Liaoning	168018	30357	6184	36888		27284	3350
吉 林 Jilin	155832	31982		6		64363	140
黑 龙 江 Heilongjiang	413777	99776	4891	70486		40965	
江 苏 Jiangsu	1273285	172842	33335	6498		256633	
浙 江 Zhejiang	1793739	104619	27286	9736	1095	772762	17120
安 徽 Anhui	2504765	189458	70529	18022		1173436	
福 建 Fujian	1543072	113310	34550			651110	25401
江 西 Jiangxi	3536160	244745	69566	200		1654110	18150
山 东 Shandong	2064093	81501	26967	219698		545288	15980
河 南 Henan	2641337	199647	44802	143219		867788	874
湖 北 Hubei	2104466	110181	69126		2130	608384	13426
湖 南 Hunan	1523533	177235	43700			587760	4953
广 东 Guangdong	408237	40252	1835			171951	
广 西 Guangxi	779493	79125	12733			462678	
海 南 Hainan	140582	11957	3742			54405	69
重 庆 Chongqing	844555	19089	8270			571077	5061
四 川 Sichuan	3953797	179002	117663	23548		1918237	32784
贵 州 Guizhou	4409225	287835	62092		3915	822358	63849
云 南 Yunnan	2007119	147247	49178		3088	729597	2000
西 藏 Tibet	77121	7989				47961	
陕 西 Shaanxi	2275206	55171	62771	142003	1540	623463	57909
甘 肃 Gansu	1479191	107991	10700	307173	1847	570866	14926
青 海 Qinghai	160961	3970		10865		88335	2565
宁 夏 Ningxia	137363	7298	500	17826		53400	2443
新 疆 Xinjiang	1268632	173779	22730	211399	3146	267914	4162

National Investment in Fixed Assets of County Seat Service Facilities by Industry(2022)

Measurement Unit: 10,000 RMB

排　水 Sewerage	污水处理 Wastewater Treatment	污泥处置 Sludge Disposal	再生水利用 Wastewater Recycled and Reused	园林绿化 Landscaping	市容环境卫生 Environmental Sanitation	垃圾处理 Domestic Garbage Treatment	其他 Other	本年新增固定资产 Newly Added Fixed Assets of This Year	地区名称 Name of Regions
7717120	3110748	56788	82576	3525081	2225108	1606255	8390367	25380488	全　国
732140	168435	3166	8445	280410	520613	469598	365328	1431874	河　北
191119	90829	303	705	132210	13946	10088	239298	783534	山　西
119230	38919	4164	6812	61587	34932	31724	114427	221004	内 蒙 古
46367	14511			2952	10289	8297	4347	54959	辽　宁
38600	4313		4100	5429	4114	1003	11198	58084	吉　林
146145	59638			7310	2960	2231	41244	179681	黑 龙 江
374401	133934	3518		128118	43001	35951	258457	232217	江　苏
277120	162230			171215	85933	54557	326853	1045995	浙　江
554791	216094	1709	4623	219774	60269	17961	218486	791702	安　徽
191987	106622	16		104124	98417	89016	324173	665615	福　建
375516	169947	7043		356626	196351	36342	620896	2277814	江　西
673406	78433	3000	280	270300	58980	38365	171973	1545576	山　东
486881	179217	6755	4428	595586	253342	232266	49198	2317669	河　南
382500	94460			123628	46540	27131	748551	1277562	湖　北
371074	254403	3100		80466	52377	48475	205968	798834	湖　南
88501	47660			3221	1646	892	100831	178134	广　东
150131	47813			53175	14903	13441	6748	425413	广　西
26168	18570			551	971	940	42719	26313	海　南
74010	25316	430		98262	26632	22428	42154	684981	重　庆
957080	476889	11967	4018	352042	179827	97940	193614	3728268	四　川
389799	243352	3200	1520	54605	196039	151237	2528733	2250413	贵　州
306777	135216	1340		183919	65785	29730	519528	1270034	云　南
6839	2787			123	5903	4570	8306	46089	西　藏
440761	185714		8000	78240	60281	38604	753067	1140200	陕　西
188684	104689	3577	10675	85988	79204	67974	111812	1005447	甘　肃
15079	7351			5110	14118	6624	20919	135053	青　海
15849	6302	500	2290	10010	12036	9448	18001	51987	宁　夏
96165	37104	3000	26680	60100	85699	59422	343538	756036	新　疆

2-5-1 按资金来源分全国历年县城市政公用设施建设固定资产投资

计量单位：亿元

年份 Year	本年资金 来源合计 Completed Investment of This Year	上 年 末 结余资金 The Balance of The revious Year	本年资金来源		
			小 计 Subtotal	中央财政 拨 款 Financial Allocation From Central Government Budget	地方财政 拨 款 Financial Allocation From Local Government Budget
2001	306.4	7.6	298.8	13.7	47.4
2002	377.1	4.2	372.9	19.7	74.2
2003	518.1	6.8	511.3	30.2	101.6
2004	619.2	7.3	611.9	20.6	133.4
2005	682.8	9.5	673.3	25.1	170.1
2006	755.2	16.9	738.3	54.8	255.4
2007	833.0	13.1	819.8	34.5	346.7
2008	1126.7	16.3	1110.4	52.5	520.4
2009	1682.9	20.8	1662.1	107.0	662.0
2010	2559.8	34.3	2525.5	325.5	985.8
2011	2872.6	31.7	2840.9	152.4	1523.8
2012	3887.7	51.7	3835.9	205.0	2073.7
2013	3683.0	60.7	3622.2	135.8	1070.7
2014	3690.4	63.2	3627.2	111.8	968.5
2015	3011.2	52.8	2958.4	121.7	930.7
2016	3190.9	30.7	3160.2	120.6	1074.4
2017	3610.8	86.6	3524.2	107.0	1035.3
2018	3222.3	102.6	3119.6	132.0	797.0
2019	3900.5	148.0	3752.5	169.3	892.6
2020	4371.0	210.6	4160.4	252.2	1135.1
2021	4929.7	323.1	4606.5	300.2	1036.9
2022	4681.0	371.4	4309.6	180.1	1024.1

注：1. 自2013年起，"本年资金来源合计"为"本年实际到位资金合计"。
2. 自2013年起，"中央财政拨款"为"中央预算资金"，"地方财政拨款"为除"中央预算资金"外的"国家预算资金"合计。

National Fixed Assets Investment of County Seat Service Facilities

Measurement Unit: 100 million RMB

Sources of Fund					
国内贷款 Domestic Loan	债　券 Securities	利用外资 Foreign Investment	自筹资金 Self- Raised Funds	其他资金 Other Funds	年份 Year
33.2	1.5	23.4	116.1	63.5	2001
46.3	1.2	12.6	149.5	69.4	2002
69.0	1.6	25.1	202.2	81.7	2003
84.1	2.2	38.7	222.1	110.7	2004
76.9	2.2	39.9	247.3	111.9	2005
89.6	1.5	26.2	234.2	76.6	2006
88.1	2.2	26.3	240.3	81.8	2007
107.6	1.4	28.0	297.0	103.5	2008
298.6	11.7	32.9	385.5	164.4	2009
332.6	4.1	44.3	606.6	226.6	2010
278.0	3.2	31.8	644.6	207.2	2011
318.1	2.3	85.0	877.2	274.7	2012
277.2	7.7	56.1	1610.4	464.3	2013
315.5	4.1	23.3	1767.7	436.3	2014
222.2	5.1	33.1	1233.3	412.2	2015
221.6	14.0	19.0	1339.4	371.3	2016
378.9	32.9	21.2	1372.3	576.5	2017
171.5	38.4	16.8	1307.4	656.4	2018
174.6	92.1	32.0	1457.7	934.1	2019
224.0	406.7	29.0	1169.9	943.6	2020
311.0	503.2	39.3	1359.8	1055.9	2021
208.2	710.8	37.6	1231.6	917.2	2022

Notes: 1.Since 2013, Completed Investment of This Year is changed to The Total Funds Actually Available for The Reported Year.
2.Since 2013, Financial Allocation from Central Government Budget is changed to Central Budgetary Fund, and Financial Allocation from Local Government Budget is State Budgetary Fund excluding Central Budgetary Fund.

2-5-2 按资金来源分全国县城市政公用设施建设固定资产投资(2022年)

计量单位: 万元

地区名称 Name of Regions	本市实际到位资金合计 The Total Funds Actually Available for The Reported Year	上年末结余资金 The Balance of The Previous Year	本年资金来源			国内贷款 Domestic Loan
			小 计 Subtotal	国家预算资金 State Budgetary Fund	中央预算资金 Central Budgetary Fund	
全 国 National Total	46809666	3713945	43095721	12041338	1800631	2082034
河 北 Hebei	3354204	52929	3301275	1072314	161713	58544
山 西 Shanxi	1243873	119704	1124169	581158	71319	18084
内 蒙 古 Inner Mongolia	684937	49853	635084	248671	33904	11390
辽 宁 Liaoning	186797	3769	183028	54690	9538	
吉 林 Jilin	195458	18694	176764	16112	1862	
黑 龙 江 Heilongjiang	468034	22635	445399	233744	45276	
江 苏 Jiangsu	1295295	9306	1285989	378408	5080	20612
浙 江 Zhejiang	1934077	108910	1825167	392277	18051	43032
安 徽 Anhui	2731910	39675	2692235	1214669	4012	49474
福 建 Fujian	1795774	90733	1705041	395041	7241	96151
江 西 Jiangxi	4453920	708655	3745265	1315231	32453	335746
山 东 Shandong	1964565	55353	1909212	735425	15768	55780
河 南 Henan	2649652	110219	2539433	1001748	54345	156508
湖 北 Hubei	2667293	154451	2512842	490980	72251	14000
湖 南 Hunan	2077851	191457	1886394	272341	55341	63100
广 东 Guangdong	466358	4879	461479	84450	5437	
广 西 Guangxi	846936	58454	788482	287125	14616	90061
海 南 Hainan	157481	29089	128392	76659	45059	
重 庆 Chongqing	875040	95074	779966	251194	24921	229154
四 川 Sichuan	4367880	580643	3787237	379415	162060	501747
贵 州 Guizhou	4382805	653702	3729103	676901	156994	144177
云 南 Yunnan	1973145	101388	1871757	488752	160898	74307
西 藏 Tibet	163209	30502	132707	105642	84707	
陕 西 Shaanxi	2351763	105709	2246054	512451	103480	61684
甘 肃 Gansu	1394618	125203	1269415	230545	131053	14040
青 海 Qinghai	347049	79484	267565	205041	112434	
宁 夏 Ningxia	95165	3519	91646	20509	14977	4583
新 疆 Xinjiang	1684577	109956	1574621	319845	195841	39860

National Investment in Fixed Assets of County Seat Service Facilities by Capital Source(2022)

Measurement Unit: 10,000 RMB

| Sources of Fund | | | | 各项应付款 | 地区名称 |
债　券 Securities	利用外资 Foreign Investment	自筹资金 Self-Raised Funds	其他资金 Other Funds	Sum Payable This Year	Name of Regions
7108112	**376350**	**12316369**	**9171518**	**6931818**	全　　国
1299093		772722	98602	461281	河　　北
132406	25000	237133	130388	396076	山　　西
149657		160162	65204	83001	内　蒙　古
101574		23897	2867	6281	辽　　宁
129151		29074	2427	14768	吉　　林
139819		45516	26320	17000	黑　龙　江
55242		314424	517303	273300	江　　苏
114785		897724	377349	63826	浙　　江
104501	1195	859563	462833	437345	安　　徽
181363		517962	514524	145131	福　　建
217362	1050	1009370	866506	449535	江　　西
323046		478174	316787	241358	山　　东
270729	408	766895	343145	350679	河　　南
192784	226700	944051	644327	611813	湖　　北
269874	9288	1067468	204323	42871	湖　　南
252751		37965	86313	60849	广　　东
57910		250576	102810	48196	广　　西
		9236	42497	56117	海　　南
69881		158419	71318	115146	重　　庆
704511	39335	469488	1692741	649021	四　　川
264496	10621	1428816	1204092	1135309	贵　　州
425585	4329	356318	522466	558599	云　　南
3524		11066	12475		西　　藏
243051	14383	993807	420678	272370	陕　　西
453219		358483	213128	226678	甘　　肃
31599		9390	21535	1056	青　　海
44761		10439	11354	33273	宁　　夏
875438	44041	98231	197206	180939	新　　疆

二、居民生活数据
Data by Residents Living

2-6-1　全国历年县城供水情况

年　份 Year	综　合 生产能力 （万立方米／日） Integrated Production Capacity (10,000 m³/day)	供水管道 长　度 （公里） Length of Water Supply Pipelines (km)	供水总量 （万立方米） Total Quantity of Water Supply (10,000 m³)	生活用量 Residential Use
2000	3662	70046	593588	310331
2001	3754	77316	577948	327081
2002	3705	78315	567915	338689
2003	3400	87359	606189	363311
2004	3680	92867	653814	395181
2005	3862	98980	676548	409045
2006	4207	113553	746892	407389
2007	5744	131541	794495	448943
2008	5976	142507	826300	459866
2009	4775	148578	856291	484644
2010	4683	159905	925705	509266
2011	5174	173452	977115	534918
2012	5446	186500	1020298	565835
2013	5239	194465	1038662	584759
2014	5437	203514	1063270	600436
2015	5769	214736	1069191	612383
2016	5421	211355	1064966	609171
2017	6443	234466	1128373	636403
2018	7415	242537	1145082	660469
2019	6304	258602	1190883	697214
2020	6451	272990	1190206	718585
2021	6945	278522	1219928	734577
2022	6920	293456	1261942	764822

注：自2006年起，供水普及率指标按城区人口和城区暂住人口合计为分母计算。

National County Seat Water Supply in Past Years

用水人口 （万人） Population with Access to Water Supply (10,000 persons)	人　均　日 生活用水量 （升） Daily Water Consumption Per Capita (liter)	供　水 普及率 （%） Water Coverage Rate (%)	年　份 Year
6931.2	122.7	84.83	2000
6889.2	130.1	76.45	2001
7145.7	129.9	80.53	2002
7532.9	132.1	81.57	2003
7931.1	136.5	82.26	2004
8342.2	134.3	83.18	2005
9093.4	122.7	76.43	2006
10218.7	120.4	81.15	2007
10628.4	119.4	81.59	2008
11200.7	118.6	83.72	2009
11811.4	118.9	85.14	2010
12345.0	118.7	86.09	2011
12971.0	119.5	86.94	2012
13456.0	119.1	88.14	2013
13913.4	118.2	88.89	2014
14048.3	119.4	89.96	2015
13974.9	119.4	90.50	2016
14508.9	120.2	92.87	2017
14722.5	122.9	93.80	2018
15082.0	126.7	95.06	2019
15316.7	128.5	96.66	2020
15252.2	132.0	97.42	2021
15275.2	137.2	97.86	2022

Note: Since 2006,water coverage rate has been calculated based on denominator which combines both permanent and temporary residents in urban areas, and the datas of brackets are the same index but calculated by the method of past years.

2-6-2 县城供水(2022年)

地区名称 Name of Regions		综 合 生产能力 (万立方 米/日) Integrated Production Capacity (10,000 m³/ day)	地下水 Underground Water	供水管道 长 度 (公里) Length of Water Supply Pipelines (km)	建成区 in Built District	供水总量 (万立方米) Total Quantity of Water Supply (10,000 m³)
全 国	National Total	6919.57	1555.68	293455.68	259369.61	1261941.93
河 北	Hebei	408.64	154.35	18079.95	17183.77	71233.67
山 西	Shanxi	171.11	118.55	10964.21	8945.65	33026.24
内 蒙 古	Inner Mongolia	160.23	157.22	12233.72	11880.52	30347.81
辽 宁	Liaoning	90.68	20.96	5465.57	4652.21	18360.82
吉 林	Jilin	49.91	13.51	3225.74	3152.92	11027.52
黑 龙 江	Heilongjiang	101.05	84.52	6077.29	5905.92	18588.11
江 苏	Jiangsu	304.56	15.48	12805.47	10710.40	57684.71
浙 江	Zhejiang	405.88	0.03	20563.49	13668.46	73164.63
安 徽	Anhui	399.19	89.75	20126.48	18634.61	88454.48
福 建	Fujian	270.21	4.89	10434.77	9402.57	50087.41
江 西	Jiangxi	456.51	16.59	20924.03	19747.48	73199.54
山 东	Shandong	487.89	204.17	13400.03	12300.74	89460.75
河 南	Henan	616.73	249.34	15243.29	14389.58	97405.54
湖 北	Hubei	247.54	6.50	9263.21	8646.05	42521.88
湖 南	Hunan	494.58	38.03	21512.42	19668.95	107117.72
广 东	Guangdong	271.60	17.18	11244.62	9047.75	51545.21
广 西	Guangxi	264.58	31.22	9601.12	9514.25	51481.40
海 南	Hainan	153.90	9.50	2245.82	988.96	8512.68
重 庆	Chongqing	77.05		2403.38	2375.53	15970.58
四 川	Sichuan	408.33	22.41	17666.68	15526.02	89433.14
贵 州	Guizhou	234.50	13.12	11864.18	10376.01	42874.63
云 南	Yunnan	278.52	8.18	12047.98	10009.44	41132.15
西 藏	Tibet	39.40	21.95	1685.00	1484.82	5555.84
陕 西	Shaanxi	164.90	85.52	6236.69	4720.62	28894.26
甘 肃	Gansu	98.25	37.50	5551.32	4980.60	18149.82
青 海	Qinghai	39.29	9.71	2147.55	2002.88	5621.26
宁 夏	Ningxia	59.38	32.71	2084.63	1934.33	8681.01
新 疆	Xinjiang	165.16	92.79	8357.04	7518.57	32409.12

County Seat Water Supply(2022)

生产运营 用　水 The Quantity of Water for Production and Operation	公共服务 用　水 The Quantity of Water for Public Service	居民家庭 用　水 The Quantity of Water for Household Use	其他用水 The Quantity of Water for Other Purposes	用水户数 （户） Number of Households with Access to Water Supply (unit)	家庭用户 Household User	用水人口 （万人） Population with Access to Water Supply (10,000 persons)	地区名称 Name of Regions
259749.93	118142.55	638901.74	62834.74	57269961	50323756	15275.21	全　　国
16253.87	5912.04	34460.36	3247.40	3774599	3413316	1047.92	河　　北
6964.89	4037.12	16951.45	1301.14	1653745	1509566	603.05	山　　西
6855.72	3737.24	14439.48	1358.21	2290543	1957086	471.25	内　蒙　古
2033.86	1456.91	7927.84	1500.28	1137291	971848	198.46	辽　　宁
1487.82	1015.52	4885.31	747.89	843208	758546	155.55	吉　　林
2539.88	1963.32	9986.77	484.00	1708774	1507919	333.54	黑　龙　江
13000.59	5785.65	27444.20	2848.03	2576615	2321298	519.55	江　　苏
21693.50	5801.73	32206.06	4251.13	2866881	2421705	458.04	浙　　江
24006.39	7566.51	42585.64	3860.29	3762387	3352451	897.11	安　　徽
7096.79	5947.39	25706.69	2354.74	1942161	1639647	465.27	福　　建
15374.97	5671.85	36990.23	4968.56	3806339	3306781	715.81	江　　西
36926.15	6986.79	34939.31	3826.70	2648125	2412816	1023.80	山　　东
20694.77	11464.41	48952.12	5287.72	4417426	4111572	1477.85	河　　南
7179.15	3376.56	22301.80	1514.25	1741144	1472929	456.73	湖　　北
17352.68	10545.81	57591.94	3578.37	3620617	3102769	1143.70	湖　　南
8679.10	3075.34	28022.92	3657.58	1748979	1509378	512.69	广　　东
9365.85	5229.89	28920.31	1126.34	1652900	1485576	515.94	广　　西
1694.05	1066.33	4300.58	308.65	313415	268136	60.84	海　　南
1808.54	1466.31	9020.65	857.18	944570	825195	212.18	重　　庆
13160.98	8031.12	47847.04	6138.52	4683439	4125749	1166.01	四　　川
4114.03	1604.90	27704.53	1670.26	2298138	1908636	651.52	贵　　州
7003.15	4250.83	22426.68	1445.36	1977950	1741166	590.16	云　　南
222.39	396.77	2936.50	223.48	231151	183908	72.10	西　　藏
5027.55	3000.45	16994.11	1005.05	1420852	1151418	506.48	陕　　西
2683.31	2495.09	10048.36	831.23	1073983	941320	402.97	甘　　肃
694.13	874.51	2836.36	299.78	319943	278247	110.17	青　　海
1727.46	1376.75	3477.84	1053.92	499294	434462	119.31	宁　　夏
4108.36	4005.41	16996.66	3088.68	1315492	1210316	387.21	新　　疆

2-6-3 县城供水(公共供水)(2022年)

地区名称 Name of Regions	综合生产能力 (万立方米/日) Integrated Production Capacity (10,000 m³/day)	地下水 Under-ground Water	水厂个数 (个) Number of Water Plants (unit)	地下水 Under-ground Water	供水管道长度 (公里) Length of Water Supply Pipelines (km)	供水总量（万立方米）合计 Total	售水量 小计 Subtotal	售水量 生产运营用水 The Quantity of Water for Production and Operation
全 国 National Total	6058.45	1072.61	2449	807	283361.20	1162156.52	979843.55	194496.87
河 北 Hebei	365.50	120.97	164	81	16970.20	65650.25	54290.25	11558.31
山 西 Shanxi	121.50	80.59	129	104	9514.19	26453.66	22682.02	3429.56
内 蒙 古 Inner Mongolia	129.67	126.92	110	107	11930.05	24528.90	20571.74	2898.12
辽 宁 Liaoning	80.48	13.26	31	11	5011.57	16428.46	10986.53	1320.31
吉 林 Jilin	45.50	9.10	20	9	3141.18	10373.63	7482.65	991.19
黑 龙 江 Heilongjiang	91.19	74.66	59	51	5886.45	17757.70	14143.56	1832.38
江 苏 Jiangsu	280.20	4.70	24		12458.42	53630.42	45024.18	10619.47
浙 江 Zhejiang	388.30		55	1	20461.47	72361.40	63149.19	21310.27
安 徽 Anhui	351.49	50.39	80	17	19360.46	76573.99	66138.34	15112.83
福 建 Fujian	261.28	4.60	71	9	10376.65	49368.15	40386.35	6565.03
江 西 Jiangxi	438.20	1.03	107	1	20749.19	73025.64	62831.71	15300.19
山 东 Shandong	360.02	110.14	122	49	12326.77	65049.89	58268.09	18044.25
河 南 Henan	509.34	159.52	151	90	14023.66	77692.11	66685.59	10250.73
湖 北 Hubei	240.14	6.00	66		9204.08	42093.16	33943.04	6970.63
湖 南 Hunan	475.00	32.50	106	8	21263.33	105695.95	87647.03	16770.15
广 东 Guangdong	255.20	9.00	58	2	11031.48	51464.67	43354.40	8635.52
广 西 Guangxi	231.20	6.67	80	9	9497.50	47983.24	41144.23	6680.99
海 南 Hainan	54.70		15		2175.72	7918.59	6775.52	1243.85
重 庆 Chongqing	77.05		34		2403.38	15970.58	13152.68	1808.54
四 川 Sichuan	388.71	13.67	188	11	17325.66	86334.25	72078.77	11132.87
贵 州 Guizhou	231.29	13.05	151	17	11853.18	42856.01	35075.10	4111.23
云 南 Yunnan	191.49	3.37	179	11	11904.15	40379.75	34373.62	6764.73
西 藏 Tibet	24.62	14.52	77	45	1477.30	5068.69	3291.99	186.70
陕 西 Shaanxi	120.75	57.02	125	58	5493.13	24283.40	21416.30	2733.99
甘 肃 Gansu	93.49	33.75	87	38	5333.52	17584.40	15492.57	2472.44
青 海 Qinghai	38.79	9.21	42	8	2140.55	5534.81	4618.33	610.68
宁 夏 Ningxia	55.40	29.73	23	6	1923.92	7963.19	6918.15	1132.07
新 疆 Xinjiang	157.95	88.24	95	64	8124.04	32131.63	27921.62	4009.84

County Seat Water Supply (Public Water Suppliers)(2022)

Total Quantity of Water Supply(10,000m³)						用水户数 （户） Number of Households with Access to Water Supply (unit)	居民家庭 Households	用水人口 （万人） Population with Access to Water Supply (10,000 persons)	地区名称 Name of Regions
Water Sold			免费供水量 The Quantity of Free Water Supply	生活用水 Domestic Water Use	漏损水量 The Lossed Water				
公共服务用水 The Quantity of Water for Public Service	居民家庭用水 The Quantity of Water for Household Use	其他用水 The Quantity of Other Purposes							
107694.17	621484.21	56168.30	36504.35	7777.59	145808.62	56092751	49396947	15038.20	全　国
5567.79	34162.97	3001.18	2074.66	129.42	9285.34	3727468	3378828	1043.93	河　北
3299.84	15107.57	845.05	742.46	252.50	3029.18	1495573	1365541	568.48	山　西
2982.64	13641.89	1049.09	693.16	184.68	3264.00	2217870	1941664	466.06	内　蒙　古
1230.83	7517.84	917.55	588.69	61.43	4853.24	1079905	919492	195.78	辽　宁
969.78	4831.67	690.01	698.89	84.96	2192.09	834905	751490	153.58	吉　林
1879.32	9980.27	451.59	818.55	127.13	2795.59	1690853	1494609	333.18	黑　龙　江
5393.43	26274.32	2736.96	1325.83	635.10	7280.41	2569356	2314583	518.78	江　苏
5381.73	32206.06	4251.13	1559.36	73.00	7652.85	2866776	2421705	458.04	浙　江
6776.17	41068.33	3181.01	1641.91	243.83	8793.74	3664501	3269974	870.48	安　徽
5883.95	25588.52	2348.85	2046.50	47.25	6935.30	1938786	1637206	464.45	福　建
5652.59	36910.48	4968.45	1560.10	337.92	8633.83	3802475	3303036	714.49	江　西
5788.77	32248.50	2186.57	428.96	110.79	6352.84	2517528	2314101	979.30	山　东
7558.79	45430.28	3445.79	2749.66	330.54	8256.86	4181562	3910483	1415.78	河　南
3352.96	22143.50	1475.95	2740.45	488.63	5409.67	1726720	1460155	454.49	湖　北
10268.70	57050.99	3557.19	5283.49	1392.75	12765.43	3602404	3084769	1138.19	湖　南
3054.76	28022.92	3641.20	498.44	97.01	7611.83	1712221	1509378	512.69	广　东
5123.94	28318.31	1020.99	944.74	95.88	5894.27	1625158	1461070	508.23	广　西
1001.13	4221.89	308.65	187.66	41.96	955.41	296894	251887	60.82	海　南
1466.31	9020.65	857.18	363.62	69.00	2454.28	944570	825195	212.18	重　庆
7730.22	47102.18	6113.50	2760.78	1069.58	11494.70	4628586	4079756	1159.83	四　川
1600.06	27695.10	1668.71	542.54	75.78	7238.37	2296714	1908335	649.28	贵　州
4200.25	22066.86	1341.78	2169.22	592.23	3836.91	1971831	1736623	585.17	云　南
349.00	2555.48	200.81	1328.66	1034.94	448.04	204570	161971	66.18	西　藏
2573.33	15217.34	891.64	603.46	72.65	2263.64	1300676	1042585	491.17	陕　西
2400.14	9900.26	719.73	450.67	28.09	1641.16	1070050	937581	401.78	甘　肃
874.51	2836.36	296.78	381.43	10.44	535.05	319943	278247	110.17	青　海
1368.45	3477.84	939.79	129.68	19.00	915.36	499277	434462	119.31	宁　夏
3964.78	16885.83	3061.17	1190.78	71.10	3019.23	1305579	1202221	386.38	新　疆

2-6-4 县城供水(自建设施供水)(2022年)

地区名称 Name of Regions	综 合 生产能力 (万立方米/日) Integrated Production Capacity (10,000 m³/day)	地下水 Underground Water	供水管道 长 度 (公里) Length of Water Supply Pipelines (km)	建成区 in Built District	供水总量(万立方米)	
					合计 Total	生产运营 用 水 The Quantity of Water for Production and Operation
全 国 National Total	861.12	483.07	10094.48	7446.07	99785.41	65253.06
河 北 Hebei	43.14	33.38	1109.75	955.73	5583.42	4695.56
山 西 Shanxi	49.61	37.96	1450.02	866.64	6572.58	3535.33
内 蒙 古 Inner Mongolia	30.56	30.30	303.67	276.87	5818.91	3957.60
辽 宁 Liaoning	10.20	7.70	454.00	422.00	1932.36	713.55
吉 林 Jilin	4.41	4.41	84.56	84.56	653.89	496.63
黑 龙 江 Heilongjiang	9.86	9.86	190.84	149.24	830.41	707.50
江 苏 Jiangsu	24.36	10.78	347.05	296.02	4054.29	2381.12
浙 江 Zhejiang	17.58	0.03	102.02	7.02	803.23	383.23
安 徽 Anhui	47.70	39.36	766.02	514.92	11880.49	8893.56
福 建 Fujian	8.93	0.29	58.12	3.00	719.26	531.76
江 西 Jiangxi	18.31	15.56	174.84	164.56	173.90	74.78
山 东 Shandong	127.87	94.03	1073.26	802.34	24410.86	18881.90
河 南 Henan	107.39	89.82	1219.63	915.47	19713.43	10444.04
湖 北 Hubei	7.40	0.50	59.13	27.00	428.72	208.52
湖 南 Hunan	19.58	5.53	249.09	231.46	1421.77	582.53
广 东 Guangdong	16.40	8.18	213.14	49.03	80.54	43.58
广 西 Guangxi	33.38	24.55	103.62	103.62	3498.16	2684.86
海 南 Hainan	99.20	9.50	70.10	70.10	594.09	450.20
重 庆 Chongqing						
四 川 Sichuan	19.62	8.74	341.02	218.05	3098.89	2028.11
贵 州 Guizhou	3.21	0.07	11.00	0.43	18.62	2.80
云 南 Yunnan	87.03	4.81	143.83	135.01	752.40	238.42
西 藏 Tibet	14.78	7.43	207.70	207.70	487.15	35.69
陕 西 Shaanxi	44.15	28.50	743.56	448.68	4610.86	2293.56
甘 肃 Gansu	4.76	3.75	217.80	129.01	565.42	210.87
青 海 Qinghai	0.50	0.50	7.00	7.00	86.45	83.45
宁 夏 Ningxia	3.98	2.98	160.71	127.61	717.82	595.39
新 疆 Xinjiang	7.21	4.55	233.00	233.00	277.49	98.52

County Seat Water Supply (Suppliers with Self-Built Facilities)(2022)

Total Quantity of Water Supply(10,000m³)			用水户数 （户）		用水人口 （万人）	地区名称
公共服务 用　　水 The Quantity of Water for Public Service	居民家庭 用　　水 The Quantity of Water for Household Use	其他用水 The Quantity of Water for Other Purposes	Number of Households with Access to Water Supply (unit)	居民家庭 Households	Population with Access to Water Supply (10,000 persons)	Name of Regions
10448.38	17417.53	6666.44	1177210	926809	237.01	全　　国
344.25	297.39	246.22	47131	34488	3.99	河　　北
737.28	1843.88	456.09	158172	144025	34.57	山　　西
754.60	797.59	309.12	72673	15422	5.19	内 蒙 古
226.08	410.00	582.73	57386	52356	2.68	辽　　宁
45.74	53.64	57.88	8303	7056	1.97	吉　　林
84.00	6.50	32.41	17921	13310	0.36	黑 龙 江
392.22	1169.88	111.07	7259	6715	0.77	江　　苏
420.00			105			浙　　江
790.34	1517.31	679.28	97886	82477	26.63	安　　徽
63.44	118.17	5.89	3375	2441	0.82	福　　建
19.26	79.75	0.11	3864	3745	1.32	江　　西
1198.02	2690.81	1640.13	130597	98715	44.50	山　　东
3905.62	3521.84	1841.93	235864	201089	62.07	河　　南
23.60	158.30	38.30	14424	12774	2.24	湖　　北
277.11	540.95	21.18	18213	18000	5.51	湖　　南
20.58		16.38	36758			广　　东
105.95	602.00	105.35	27742	24506	7.71	广　　西
65.20	78.69		16521	16249	0.02	海　　南
						重　　庆
300.90	744.86	25.02	54853	45993	6.18	四　　川
4.84	9.43	1.55	1424	301	2.24	贵　　州
50.58	359.82	103.58	6119	4543	4.99	云　　南
47.77	381.02	22.67	26581	21937	5.92	西　　藏
427.12	1776.77	113.41	120176	108833	15.31	陕　　西
94.95	148.10	111.50	3933	3739	1.19	甘　　肃
		3.00				青　　海
8.30		114.13	17			宁　　夏
40.63	110.83	27.51	9913	8095	0.83	新　　疆

2-7-1　全国历年县城节约用水情况

年 份 Year	计划用水量 （万立方米） Planned Quantity of Water Use (10,000 m³)	新水取用量 （万立方米） Fresh Water Used (10,000 m³)
2000	115591	109069
2001	161795	133634
2002	110702	91610
2003	120551	94938
2004	123176	98004
2005	155933	118727
2006		122700
2007		127318
2008		122275
2009		139735
2010		157563
2011		119923
2012		180603
2013		136223
2014		140966
2015		126923
2016		115707
2017		106174
2018		111385
2019		140990
2020		188350
2021		163019
2022		176670

注：自2006年起，不统计计划用水量指标。

National County Seat Water Conservation in Past Years

工业用水重复利用量（万立方米）Quantity of Industrial Water Recycled (10,000 m³)	节约用水量（万立方米）Water Saved (10,000 m³)	年 份 Year
95763	20322	2000
24206	28161	2001
26321	19092	2002
24815	25613	2003
25790	25163	2004
32852	37206	2005
32523	21546	2006
99648	22675	2007
53278	20480	2008
130864	23135	2009
133337	35963	2010
95932	26165	2011
52051	30853	2012
159152	22286	2013
164833	30029	2014
173337	32419	2015
165369	23086	2016
167080	24873	2017
189581	26618	2018
169218	28609	2019
104108	48132	2020
126908	49765	2021
162746	54399	2022

Note: From 2006, "Planned Quantity of Water Use" has not been counted.

2-7-2 县城节约用水(2022年)

计量单位：万立方米

地区名称 Name of Regions	计划用水户数 (户) Planned Water Consumers (households)	自备水计划用水户数 Planned Self-produced Water Consumers	计划用水户实际用水量		新水取水量 Fresh Water Used	工 业 Industry	重复利用量 Water Reused
			合 计 Total	工 业 Industry			
全 国 National Total	**1137132**	**39502**	**359206**	**225927**	**176670**	**63181**	**182537**
河 北 Hebei							
山 西 Shanxi	223410	4972	12049	4123	10284	2991	1765
内 蒙 古 Inner Mongolia	127355	17	16976	7308	14804	5679	2172
辽 宁 Liaoning	87	75	947	312	842	208	104
吉 林 Jilin	232	120	1512	325	1326	149	186
黑 龙 江 Heilongjiang	16766	16	370	167	348	145	22
江 苏 Jiangsu	1672	155	41080	28009	16931	6310	24149
浙 江 Zhejiang	3378	309	27244	21648	11778	7993	15466
安 徽 Anhui	13	13	165	134	96	72	70
福 建 Fujian	9998		660	0	660	0	
江 西 Jiangxi	1028	74	9346	4308	7747	3649	1599
山 东 Shandong	5117	1244	94600	78232	23098	10860	71502
河 南 Henan	36586	11305	23394	10830	18068	7105	5326
湖 北 Hubei	1200		2470	44	2290	44	180
湖 南 Hunan	27111	4710	7415	3168	4926	2612	2489
广 东 Guangdong	81824	1002	7841	340	7694	299	147
广 西 Guangxi							
海 南 Hainan	3774	49	5396	2800	5386	2790	10
重 庆 Chongqing	371	6	1865	775	1772	687	92
四 川 Sichuan	331763	299	10995	3437	9815	2629	1180
贵 州 Guizhou	37504	430	31672	20577	8362	1187	23310
云 南 Yunnan	84279	5399	14439	1507	13607	947	833
西 藏 Tibet	201	200	15		15		
陕 西 Shaanxi	81033	7076	43300	36398	11831	5346	31469
甘 肃 Gansu	57821	2004	2464	235	2001	230	463
青 海 Qinghai							
宁 夏 Ningxia	2149	17	1576	1093	1576	1093	
新 疆 Xinjiang	2460	10	1415	155	1412	155	3

County Seat Water Conservation(2022)

Measurement Unit:10,000m³

Actual Quantity of Water Used				节约用水量		节水措施 投资总额 （万元） Total Investment in Water-Saving Measures (10,000 RMB)	地区名称 Name of Regions
工 业 Industry	超计划定额 用 水 量 Water Quantity Consumed in Excess of Quota	重复利用率 (%) Reuse Rate (%)	工 业 Industry	Water Saved	工 业 Industry		
162746	1791	50.82	72.03	54399	41234	90779	全 国
							河 北
1132	112	14.65	27.46	766	328	16927	山 西
1629	302	12.79	22.29	453	44	17925	内 蒙 古
104		11.01	33.38	116	116	383	辽 宁
177		12.28	54.33	186	117	416	吉 林
22		5.83	12.94	22	6		黑 龙 江
21699		58.79	77.47	16132	13638	21456	江 苏
13656	61	56.77	63.08	2155	1569	1271	浙 江
62		42.15	46.39	71	71	99	安 徽
	32			236			福 建
659		17.11	15.29	1085	320	498	江 西
67372	58	75.58	86.12	10557	6795	12239	山 东
3725	915	22.77	34.39	5389	3608	3048	河 南
		7.29		3			湖 北
556		33.56	17.54	877	477	2860	湖 南
41	0	1.88	12.18	37	30	182	广 东
							广 西
10		0.19	0.36	52	5	121	海 南
88	48	4.95	11.31	523	149	3555	重 庆
808	38	10.74	23.51	1373	298	428	四 川
19390	50	73.60	94.23	11232	11156	345	贵 州
560	96	5.77	37.18	423	216	1028	云 南
							西 藏
31051	63	72.68	85.31	2385	2147	7306	陕 西
6	6	18.79	2.34	67	7	93	甘 肃
							青 海
	8			197	138	600	宁 夏
	2	0.21		63			新 疆

2-8-1 全国历年县城燃气情况
National County Seat Gas in Past Years

年 份 Year	人工煤气 Man-Made Coal Gas				天然气 Natural Gas			
	供气总量（亿立方米）Total Gas Supplied (100 million m³)	居民家庭 Households	用气人口（万人）Population with Access to Gas (10,000 persons)	管道长度（公里）Length of Gas Supply Pipeline (km)	供气总量（亿立方米）Total Gas Supplied (100 million m³)	居民家庭 Households	用气人口（万人）Population with Access to Gas (10,000 persons)	管道长度（公里）Length of Gas Supply Pipeline (km)
2000	1.72	1.63	73.10	615	3.31	2.25	237	5268
2001	2.14	1.85	89.85	424	4.37	2.71	274	6400
2002	1.19	1.13	68.30	493	6.36	3.30	316	7398
2003	0.73	0.60	51.33	519	7.67	4.16	361	8397
2004	1.80	1.51	81.91	551	10.97	5.43	437	9881
2005	3.04	2.01	127.35	830	18.12	5.83	519	12602
2006	1.26	0.49	54.58	745	16.47	7.11	780	17487
2007	1.44	0.51	58.52	1158	24.45	7.01	943	21882
2008	2.68	1.84	72.22	1426	23.26	9.00	1123	27110
2009	1.78	0.97	69.12	1459	32.16	13.79	1404	34214
2010	4.06	1.03	69.97	1520	39.98	17.09	1835	42156
2011	9.51	1.14	66.53	1458	53.87	22.83	2414	52450
2012	8.57	1.08	53.58	1255	70.14	28.11	2926	66697
2013	7.65	2.24	63.15	1345	81.58	31.55	3555	77122
2014	8.49	2.12	56.05	1542	92.65	34.50	4165	88862
2015	8.21	2.29	55.50	1376	102.60	37.97	4715	106466
2016	7.18	1.03	60.08	1366	105.70	39.25	5096	105966
2017	7.41	1.38	65.65	1296	137.96	48.00	6186	126539
2018	6.22	3.31	68.90	1869	171.04	57.61	6848	144314
2019	3.62	1.69	70.20	2671	201.87	67.21	7520	167650
2020	4.17	1.92	69.50	2761	214.53	72.13	8170	186244
2021	4.42	2.23	69.91	2743	253.64	79.82	8679	206014
2022	3.96	1.90	57.01	2289	273.64	87.15	9194	225873

2-8-1 续表 continued

年 份 Year	液化石油气　LPG				燃气普及率 (%) Gas Coverage Rate
	供气总量 （万吨） Total Gas Supplied (10,000 tons)	居民家庭 Households	用气人口 （万人） Population with Access to Gas (10,000 persons)	管道长度 （公里） Length of Gas Supply Pipeline (km)	(%)
2000	110.84	98.23	2723	96	54.41
2001	127.55	109.90	3653	674	44.55
2002	142.42	124.23	4025	760	49.69
2003	174.45	140.17	4508	1033	53.28
2004	188.94	156.37	4961	1172	56.87
2005	185.90	147.10	5151	1203	57.80
2006	195.04	147.07	5405	1728	52.45
2007	203.22	158.60	6217	2355	57.33
2008	202.14	160.60	6504	2899	59.11
2009	212.58	171.00	6776	3136	61.66
2010	218.50	174.97	7098	3053	64.89
2011	242.17	205.23	7058	2594	66.52
2012	256.94	212.13	7241	2773	68.50
2013	241.07	196.89	7208	2236	70.91
2014	235.32	198.38	7242	2538	73.24
2015	230.01	192.99	7081	2068	75.90
2016	219.22	184.36	6919	1579	78.19
2017	215.48	183.50	6458	1504	81.35
2018	214.06	181.31	6243	1829	83.85
2019	217.10	177.81	6128	1865	86.47
2020	199.54	170.15	5874	1461	89.07
2021	190.75	163.47	5391	2317	90.32
2022	187.38	156.61	5013	1386	91.38

2-8-2 县城人工煤气(2022年)

地区名称 Name of Regions		生产能力 (万立方米／日) Production Capacity (10,000 m³/day)	储气能力 (万立方米) Gas Storage Capacity (10,000 m³)	供气管道 长　度 (公里) Length of Gas Supply Pipeline (km)	自制气量 (万立方米) Self-Produced Gas (10,000 m³)	合　计 Total
全　国	National Total	35.00	341.26	2288.93	3000	39626.85
河　北	Hebei		7.00	729.18		14593.82
山　西	Shanxi	35.00	21.00	1559.75	3000	23222.00
内 蒙 古	Inner Mongolia		117.00			328.50
辽　宁	Liaoning					
吉　林	Jilin					
黑 龙 江	Heilongjiang					
江　苏	Jiangsu					
浙　江	Zhejiang					
安　徽	Anhui					
福　建	Fujian					
江　西	Jiangxi					
山　东	Shandong		3.00			1478.30
河　南	Henan					
湖　北	Hubei					
湖　南	Hunan					
广　东	Guangdong					
广　西	Guangxi					
海　南	Hainan					
重　庆	Chongqing					
四　川	Sichuan					
贵　州	Guizhou					
云　南	Yunnan		193.26			0.43
西　藏	Tibet					
陕　西	Shaanxi					3.80
甘　肃	Gansu					
青　海	Qinghai					
宁　夏	Ningxia					
新　疆	Xinjiang					

County Seat Man-Made Coal Gas(2022)

| 供气总量(万立方米) Total Gas Supplied (10,000 m³) | | 燃气损失量 | 用气户数 (户) | 居民家庭 | 用气人口 (万人) | 地区名称 |
销售气量 Quantity Sold	居民家庭 Households	燃气损失量 Loss Amount	Number of Households with Access to Gas (unit)	Households	Population with Access to Gas (10,000 persons)	Name of Regions
39007.52	**19029.32**	**619.33**	**257553**	**253253**	**57.01**	**全 国**
14555.30	4025.00	38.52	74940	74115	19.72	河 北
22685.00	14701.00	537.00	168595	166283	30.60	山 西
300.00	300.00	28.50	1300	1300	0.33	内 蒙 古
						辽 宁
						吉 林
						黑 龙 江
						江 苏
						浙 江
						安 徽
						福 建
						江 西
1463.30		15.00	30			山 东
						河 南
						湖 北
						湖 南
						广 东
						广 西
						海 南
						重 庆
						四 川
						贵 州
0.42	0.12	0.01	7568	6555	2.66	云 南
						西 藏
3.50	3.20	0.30	5120	5000	3.70	陕 西
						甘 肃
						青 海
						宁 夏
						新 疆

2-8-3 县城天然气(2022年)

地区名称 Name of Regions	储气能力 (万立方米) Gas Storage Capacity (10,000 m³)	供气管道 长 度 (公里) Length of Gas Supply Pipeline (km)	供气总量(万立方米)			
			合 计 Total	销售气量 Quantity Sold	居民家庭 Households	集中供热 Central Heating
全 国 National Total	16423.66	225872.62	2736371.24	2696642.58	871462.28	169609.30
河 北 Hebei	1117.95	26323.35	297516.33	292865.46	107175.89	31903.77
山 西 Shanxi	462.25	12177.26	199752.99	196940.80	55762.23	19525.80
内 蒙 古 Inner Mongolia	1014.67	4023.43	68039.25	66885.42	14448.82	5641.42
辽 宁 Liaoning	198.33	3188.10	66649.34	66085.55	6236.09	242.09
吉 林 Jilin	353.79	1493.14	11429.99	11333.21	3980.81	286.44
黑 龙 江 Heilongjiang	223.76	1908.77	12905.72	12675.08	4624.87	1302.90
江 苏 Jiangsu	531.85	10128.06	106735.31	105626.16	35160.41	300.00
浙 江 Zhejiang	712.90	9448.05	165678.50	164748.85	12662.31	
安 徽 Anhui	685.46	14552.60	179518.85	175531.91	42472.26	
福 建 Fujian	815.61	5562.13	55746.90	55143.21	7489.10	574.38
江 西 Jiangxi	1035.69	9906.72	127859.94	126600.10	25714.10	1073.48
山 东 Shandong	798.33	21192.96	282186.67	279455.06	71867.51	13985.04
河 南 Henan	705.83	15722.52	190899.39	187645.44	79770.06	5571.27
湖 北 Hubei	324.08	11017.69	51714.25	50841.63	22671.18	80.00
湖 南 Hunan	1970.14	12797.09	90671.52	89657.72	44105.04	80.00
广 东 Guangdong	483.32	4386.73	44606.16	44354.52	8606.70	364.00
广 西 Guangxi	288.12	2597.47	20350.71	20108.85	5513.41	
海 南 Hainan	123.95	806.11	5826.13	5779.59	657.78	
重 庆 Chongqing	67.63	2966.81	31580.83	30766.90	20155.46	429.15
四 川 Sichuan	249.85	24829.19	229428.60	222335.51	132266.89	73.13
贵 州 Guizhou	561.00	4705.53	47514.84	47153.52	9626.77	24.83
云 南 Yunnan	602.45	3802.85	25685.34	25430.33	5682.99	546.73
西 藏 Tibet						
陕 西 Shaanxi	1993.06	10634.53	144352.39	142758.80	59790.26	38416.87
甘 肃 Gansu	399.73	2447.42	55690.01	54971.80	17934.49	9320.63
青 海 Qinghai	51.30	857.11	32214.91	31788.28	11696.57	9519.05
宁 夏 Ningxia	41.82	2185.65	44890.17	44432.65	15238.85	3210.81
新 疆 Xinjiang	610.79	6211.35	146926.20	144726.23	50151.43	27137.51

County Seat Natural Gas(2022)

Total Gas Supplied (10,000 m³)		用气户数 (户) Number of Household with Access to Gas (unit)	居民家庭 Households	用气人口 (万人) Population with Access to Gas (10,000 persons)	天 然 气 汽车加气站 (座) Gas Stations for CNG-Fueled Motor Vehicles (unit)	地区名称 Name of Regions
燃气汽车 Gas-Powered Automobiles	燃气损失量 Loss Amount					
319356.07	39728.66	33548522	32324844	9193.61	1844	全　国
21099.79	4650.87	3185583	3040924	834.54	140	河　北
50686.64	2812.19	1434765	1401316	416.82	109	山　西
25062.27	1153.83	691144	644095	194.32	166	内　蒙　古
5116.44	563.79	408582	400603	103.39	56	辽　宁
4942.84	96.78	347720	336476	81.33	31	吉　林
4376.08	230.64	357437	340614	93.96	73	黑　龙　江
4255.46	1109.15	1620572	1604667	415.80	39	江　苏
982.36	929.65	953490	943010	257.17	11	浙　江
14585.08	3986.94	2280842	2140174	650.20	69	安　徽
705.22	603.69	714974	705458	216.79	8	福　建
1287.85	1259.84	1309874	1291341	341.21	10	江　西
10974.90	2731.61	3137896	3074337	889.46	120	山　东
11394.51	3253.95	3197128	3115767	1020.39	117	河　南
3519.32	872.62	1102951	1086645	282.59	37	湖　北
4843.80	1013.80	1989837	1955022	572.94	37	湖　南
9248.61	251.64	601165	597595	190.97	7	广　东
26.99	241.86	348349	343413	111.20	4	广　西
3537.01	46.54	96036	95764	18.04	9	海　南
3156.53	813.93	829063	810220	189.62	16	重　庆
16533.04	7093.09	4446747	4254144	1004.64	96	四　川
2408.17	361.32	484541	445760	172.12	36	贵　州
1950.30	255.01	277716	269571	125.88	20	云　南
						西　藏
22838.41	1593.59	1412675	1306040	374.89	168	陕　西
20093.41	718.21	564044	521229	182.46	101	甘　肃
1859.77	426.63	175328	160434	50.46	15	青　海
15656.90	457.52	369052	340138	68.69	83	宁　夏
58214.37	2199.97	1211011	1100087	333.73	266	新　疆

2-8-4 县城液化石油气(2022年)

地区名称 Name of Regions	储气能力 (吨) Gas Storage Capacity (ton)	供气管道长度 (公里) Length of Gas Supply Pipeline (km)	供气总量(吨) 合计 Total	销售气量 Quantity Sold	居民家庭 Households
全 国 National Total	**312340.93**	**1385.50**	**1873792.49**	**1860305.24**	**1566110.43**
河 北 Hebei	11116.94	157.50	69894.21	69326.36	56609.88
山 西 Shanxi	6021.35	56.14	23420.66	23018.15	15653.27
内 蒙 古 Inner Mongolia	8604.89	0.57	39919.44	39593.32	35547.54
辽 宁 Liaoning	5558.00	30.85	29755.88	29585.83	26151.75
吉 林 Jilin	2260.00	71.22	17246.40	17173.13	13731.60
黑 龙 江 Heilongjiang	5664.25	9.60	27723.89	27417.25	23856.06
江 苏 Jiangsu	16916.08	396.95	92842.90	92419.34	66462.13
浙 江 Zhejiang	10981.56	75.22	150099.33	149445.84	124531.79
安 徽 Anhui	17913.00	0.10	95000.98	94316.94	81249.10
福 建 Fujian	10975.03	32.02	84915.20	84487.80	75869.29
江 西 Jiangxi	19338.40		138788.07	137321.58	126029.58
山 东 Shandong	13955.26	96.23	71013.86	70542.57	53565.64
河 南 Henan	14881.55	1.55	138626.30	136594.77	119695.98
湖 北 Hubei	7993.65	19.37	56101.53	55593.02	49813.20
湖 南 Hunan	42902.30		182226.02	181313.39	155345.05
广 东 Guangdong	19169.05	201.00	205585.10	205274.84	162626.49
广 西 Guangxi	11167.07	8.50	137031.69	135747.41	127366.80
海 南 Hainan	1177.70		29169.16	28884.71	26001.44
重 庆 Chongqing	5301.00		12685.50	12584.30	9992.87
四 川 Sichuan	10227.28	4.33	27256.94	26920.18	20973.07
贵 州 Guizhou	17639.00	26.49	59238.47	59101.71	48746.67
云 南 Yunnan	19625.74	104.48	61096.51	60618.22	47576.77
西 藏 Tibet	2662.70	1.33	14649.05	14599.03	12663.61
陕 西 Shaanxi	14529.00	26.55	36490.57	36206.83	29549.44
甘 肃 Gansu	4262.30		25573.21	25256.79	23216.31
青 海 Qinghai	2429.82	0.08	4771.17	4677.94	3691.80
宁 夏 Ningxia	4675.51	3.00	13945.76	13770.10	7708.10
新 疆 Xinjiang	4392.50	62.42	28724.69	28513.89	21885.20

County Seat LPG Supply (2022)

Total Gas Supplied (ton)		用气户数 （户） Number of Household with Access to Gas (unit)	居民家庭 Households	用气人口 （万人） Population with Access to Gas (10,000 persons)	液化石油气 汽车加气站 （座） Gas Stations for LPG- Fueled Motor Vehicles (unit)	地区名称 Name of Regions
燃气汽车 Gas-Powered Automobiles	燃气损失量 Loss Amount					
25759.93	13487.25	15995265	15173759	5012.72	190	全　　国
535.41	567.85	653087	580442	182.90	4	河　　北
	402.51	199912	191752	70.16	2	山　　西
396.00	326.12	800702	757684	246.92	24	内 蒙 古
1099.60	170.05	371278	353599	77.58	11	辽　　宁
744.00	73.27	174964	163751	59.84	5	吉　　林
2165.60	306.64	405266	376516	118.61	15	黑 龙 江
1066.50	423.56	509710	491574	103.74	7	江　　苏
	653.49	864774	752748	200.87		浙　　江
300.00	684.04	784143	744966	238.15	2	安　　徽
	427.40	760389	729557	243.82		福　　建
	1466.49	1134469	1116580	368.61		江　　西
	471.29	419222	397858	126.87		山　　东
5525.00	2031.53	1058760	1024803	397.14	21	河　　南
3776.00	508.51	532011	481009	174.19	5	湖　　北
	912.63	1616322	1575773	527.18		湖　　南
	310.26	1179028	1146195	315.41	1	广　　东
	1284.28	1202676	1170768	402.80		广　　西
425.00	284.45	183604	179978	41.04	3	海　　南
	101.20	89365	78856	19.84	2	重　　庆
2001.09	336.76	231435	213071	90.23	12	四　　川
	136.76	991193	953920	382.37	3	贵　　州
150.00	478.29	670171	621254	246.75	1	云　　南
636.09	50.02	130026	120101	49.79	7	西　　藏
1032.89	283.74	298947	279869	97.59	8	陕　　西
813.00	316.42	358221	333810	138.05	12	甘　　肃
2.32	93.23	54016	43728	22.60	1	青　　海
960.00	175.66	74180	69319	24.01	2	宁　　夏
4131.43	210.80	247394	224278	45.66	42	新　　疆

2-9-1　全国历年县城集中供热情况

年 份 Year	供 热 能 力 Heating Capacity		供 热 总 量 Total Heat Supplied	
	蒸 汽 (吨／小时) Steam (ton/hour)	热 水 (兆瓦) Hot Water (mega watts)	蒸 汽 (万吉焦) Steam (10,000 GJ)	热 水 (万吉焦) Hot Water (10,000 GJ)
2000	4418	11548	1409	9076
2001	3647	11180	1872	20568
2002	4848	14103	2627	27245
2003	5283	19446	2871	22286
2004	5524	20891	3194	21847
2005	8837	20835	8781	18736
2006	9193	26917	8520	23535
2007	13461	35794	15780	39071
2008	12370	44082	14708	75161
2009	16675	62330	12519	56013
2010	15091	68858	16729	103005
2011	14738	81348	8475	63264
2012	13914	97281	8358	51943
2013	13285	107498	6413	68193
2014	13011	129447	5693	63928
2015	13680	125788	10957	96566
2016	10206	130430	5130	67488
2017	14853	137222	8106	148237
2018	16790	139943	8585	76441
2019	17466	153330	9810	78592
2020	18085	158186	9958	84808
2021	18646	159227	9564	87477
2022	21185	161207	10825	89735

注：2000年蒸汽供热总量计量单位为万吨。

National County Seat Centralized Heating in Past Years

管道长度 Length of Pipelines		集中供热 面　积 (亿平方米) Heated Area (100 million m²)	年　份 Year
蒸　汽 (公里) Steam (km)	热　水 (公里) Hot Water (km)		
1144	4187	0.67	2000
658	4478	0.92	2001
881	4778	1.45	2002
904	6136	1.73	2003
991	7094	1.72	2004
1176	8048	2.06	2005
1367	9450	2.37	2006
1564	12795	3.17	2007
1612	14799	3.74	2008
1874	18899	4.81	2009
1773	23737	6.09	2010
1665	28577	7.81	2011
1965	31901	9.05	2012
2929	37169	10.33	2013
2733	41209	11.42	2014
3283	43013	12.31	2015
2767	44168	13.12	2016
60793		14.63	2017
66831		16.18	2018
75068		17.48	2019
81366		18.57	2020
89274		19.45	2021
97240		20.86	2022

Note: Heating capacity through steam in 2000 is measured with the unit of 10,000 tons.

2-9-2 县城集中供热(2022年)

地区名称 Name of Regions	蒸汽 Steam						供热能力 (兆瓦) Heating Capacity (mega watts)	热电厂 供 热 Heating by Co- Generation
	供热能力 (吨/小时) Heating Capacity (ton/hour)	热电厂 供 热 Heating by Co- Generation	锅炉房 供 热 Heating by Boilers	供热总量 (万吉焦) Total Heat Supplied (10,000 GJ)	热电厂 供 热 Heating by Co- Generation	锅炉房 供 热 Heating by Boilers		
全 国 National Total	21185	18854	1817	10825	9525	983	161207	62649
河 北 Hebei	1791	1431	360	909	652	257	28350	6041
山 西 Shanxi	3963	3781	182	1824	1684	140	16795	6451
内 蒙 古 Inner Mongolia	2466	2391	75	1205	1106	98	29215	14922
辽 宁 Liaoning	1210	760	450	371	315	56	14120	9674
吉 林 Jilin				221	221		6268	494
黑 龙 江 Heilongjiang	3638	3373	265	2267	2087	180	12482	5841
江 苏 Jiangsu	135			14			11	
浙 江 Zhejiang								
安 徽 Anhui	260	260		53	53		90	90
福 建 Fujian								
江 西 Jiangxi								
山 东 Shandong	5577	5281	276	2767	2592	175	13553	10630
河 南 Henan	855	467	29	548	233	19	3687	2266
湖 北 Hubei								
湖 南 Hunan								
广 东 Guangdong								
广 西 Guangxi								
海 南 Hainan								
重 庆 Chongqing								
四 川 Sichuan	60		60	24		20	119	36
贵 州 Guizhou								
云 南 Yunnan								
西 藏 Tibet	50		50	12		11	361	231
陕 西 Shaanxi	410	410		250	250		4848	1149
甘 肃 Gansu	30		30	10		10	12053	1793
青 海 Qinghai							2278	349
宁 夏 Ningxia	740	700	40	350	332	18	3230	682
新 疆 Xinjiang							13747	2003

County Seat Central Heating(2022)

热水　Hot Water				管道长度 (公里) Length of Pipelines (km)	一级管网 First Class	二级管网 Second Class	供热面积 (万平方米) Heated Area (10,000 m²)	住宅 Housing	公共建筑 Public Building	地区名称 Name of Regions
锅炉房 供热 Heating By Boilers	供热总量 (万吉焦) Total Heat Supplied (10,000 GJ)	热电厂 供热 Heating by Co- Generation	锅炉房 供热 Heating by Boilers							
80052	**89735**	**28901**	**52251**	**97240**	**33189**	**64052**	**208596**	**155990**	**41050**	全　　国
13427	14396	3307	9350	15872	5626	10246	35192	28025	4882	河　　北
8103	13373	4373	5703	12813	3947	8866	27912	21279	5033	山　　西
12905	15133	5320	9059	14181	5004	9177	28912	19460	7232	内 蒙 古
4446	4169	1333	2767	6108	1301	4807	10001	7287	2246	辽　　宁
5757	4426	468	3918	5438	1448	3990	9458	6954	2302	吉　　林
6036	8193	3802	4089	6919	1965	4954	18459	13796	4510	黑 龙 江
11	100		2	171	156	15	249	244	4	江　　苏
										浙　　江
	9	9		171	48	124	52	29	23	安　　徽
										福　　建
										江　　西
1091	7655	5808	808	16292	6193	10099	31040	27110	2944	山　　东
471	1330	597	386	2637	1069	1569	4975	3936	532	河　　南
										湖　　北
										湖　　南
										广　　东
										广　　西
										海　　南
										重　　庆
59	61	12	23	234	82	152	165	70	81	四　　川
										贵　　州
										云　　南
61	384	114	231	824	207	616	442	154	134	西　　藏
2445	2557	940	1212	1358	705	653	6327	3858	1393	陕　　西
10229	7162	663	6426	5182	2080	3102	14464	10640	3467	甘　　肃
1914	1168	105	1036	1197	570	627	2227	1278	892	青　　海
2509	2246	709	1537	1739	767	972	5055	3703	1078	宁　　夏
10590	7376	1341	5705	6104	2022	4082	13666	8165	4295	新　　疆

三、居民出行数据
Data by Residents Travel

2-10-1 全国历年县城道路和桥梁情况
National County Seat Road and Bridge in Past Years

年 份 Year	道路长度 (万公里) Length of Roads (10,000 km)	道路面积 (亿平方米) Surface Area of Roads (100 million m²)	防洪堤长度 (万公里) Length of Flood Control Dikes (10,000 km)	人均城市道路面积 (平方米) Urban Road Surface Area Per Capita (m²)
2000	5.04	6.24	0.93	11.20
2001	5.10	7.67	0.90	8.51
2002	5.32	8.31	0.96	9.37
2003	5.77	9.06	0.93	9.82
2004	6.24	9.92	1.00	10.30
2005	6.68	10.83	0.98	10.80
2006	7.36	12.26	1.32	10.30
2007	8.38	13.44	1.17	10.70
2008	8.88	14.60	1.33	11.21
2009	9.50	15.98	1.19	11.95
2010	10.59	17.60	1.25	12.68
2011	10.86	19.24	1.37	13.42
2012	11.80	21.02	1.38	14.09
2013	12.52	22.69		14.86
2014	13.04	24.08		15.39
2015	13.35	24.95		15.98
2016	13.16	25.35		16.41
2017	14.08	26.84		17.18
2018	14.48	27.82		17.73
2019	15.16	29.01		18.29
2020	15.94	29.97		18.92
2021	16.37	30.82		19.68
2022	16.80	31.70		20.31

注：1.自2006年起，人均道路面积按县城人口和县城暂住人口合计为分母计算。
　　2.自2013年起,不再统计防洪堤长度数据。

Notes: 1.Since 2006, road surface per capita has been calculated based on denominator which combines both permanent and temporary residents in county seat areas.
　　　2.Starting from 2013, the data on the length of flood prevention dike has been unavailable.

2-10-2　县城道路和桥梁(2022年)
County Seat Roads and Bridges(2022)

地区名称 Name of Regions	道路长度 (公里) Length of Roads (km)	建成区 in Built District	道路面积 (万平方米) Surface Area of Roads (10,000 m²)	人行道 面积 Surface Area of Sidewalks	建成区 in Built District	桥梁数 (座) Number of Bridges (unit)
全　国　National Total	**168020.68**	**151399.29**	**317017.42**	**78832.58**	**291060.55**	**18344**
河　北　Hebei	13241.71	12793.86	26674.83	7311.42	25830.16	774
山　西　Shanxi	5631.09	5439.02	10831.07	2636.40	10340.09	539
内 蒙 古　Inner Mongolia	7520.86	7104.58	16197.02	4477.60	15393.78	351
辽　宁　Liaoning	1951.72	1776.07	3304.78	815.11	3013.57	196
吉　林　Jilin	1580.40	1482.75	2856.64	794.07	2756.05	156
黑 龙 江　Heilongjiang	4097.92	3959.46	5113.59	1115.07	4810.23	304
江　苏　Jiangsu	5726.27	4875.41	11879.71	2467.68	9907.70	791
浙　江　Zhejiang	6998.32	6059.58	11737.86	2636.63	10531.06	1557
安　徽　Anhui	10020.68	8823.26	23098.71	5444.54	20314.73	1497
福　建　Fujian	5724.08	4967.09	9277.26	2110.25	8198.47	636
江　西　Jiangxi	9500.67	8611.14	18465.55	4366.77	16655.04	767
山　东　Shandong	11375.23	9788.46	23658.33	4822.64	21109.73	1729
河　南　Henan	12229.16	11060.27	28260.69	6702.76	25615.93	1756
湖　北　Hubei	4471.49	4179.66	9092.54	2443.94	8572.64	528
湖　南　Hunan	11029.52	9099.79	17259.45	4731.07	16153.62	707
广　东　Guangdong	4966.90	4224.61	7710.93	1976.28	6841.15	366
广　西　Guangxi	6361.07	6180.47	11288.81	2684.47	10900.22	778
海　南　Hainan	1139.14	849.45	2104.26	583.78	1776.07	92
重　庆　Chongqing	1392.39	1366.56	2522.20	761.60	2517.86	311
四　川　Sichuan	9322.15	8618.56	18238.54	5159.05	17244.07	1127
贵　州　Guizhou	8754.31	7818.68	14243.03	3585.75	12730.90	837
云　南　Yunnan	6598.63	6035.61	11917.02	3136.96	11456.45	783
西　藏　Tibet	1158.41	831.72	1466.46	411.73	961.59	143
陕　西　Shaanxi	4961.99	4563.54	8435.13	2456.25	7861.46	567
甘　肃　Gansu	3560.09	3115.90	6444.68	1740.75	5743.02	519
青　海　Qinghai	1645.39	1525.98	2655.08	688.57	2555.70	164
宁　夏　Ningxia	1411.20	1330.98	2953.11	946.08	2817.08	87
新　疆　Xinjiang	5649.89	4916.83	9330.14	1825.36	8452.18	282

2-10-2 续表 continued

地区名称 Name of Regions	大桥及 特大桥 Great Bridge and Grand Bridge	立交桥 Intersection	道路照明 灯盏数 （盏） Number of Road Lamps (unit)	安装路灯 道路长度 （公里） Length of The Road with Street Lamp (km)	地下综合 管廊长度 （公里） Length of The Utility Tunnel (km)	新建地下综 合管廊长度 （公里） Length of The New-built Utility Tunnel (km)
全 国 National Total	**2272**	**438**	**9849652**	**126255**	**1224.56**	**355.91**
河 北 Hebei	100	58	620150	9585	89.20	16.27
山 西 Shanxi	118	23	325590	3874	19.84	22.01
内 蒙 古 Inner Mongolia	51	15	413698	5202	30.10	
辽 宁 Liaoning	6	3	135679	1463	3.20	3.40
吉 林 Jilin	37	10	149090	1143		
黑 龙 江 Heilongjiang	25	20	191266	2524		
江 苏 Jiangsu	92	14	451264	5081	6.25	
浙 江 Zhejiang	94	31	353309	5522	3.13	2.13
安 徽 Anhui	71	26	609382	8516	99.70	
福 建 Fujian	96	12	397217	4129	113.42	5.13
江 西 Jiangxi	169	28	646223	8056	13.98	1.35
山 东 Shandong	54	34	615316	8301	52.36	5.62
河 南 Henan	54	28	616036	8952	9.65	2.95
湖 北 Hubei	102	10	232914	3984	40.19	10.25
湖 南 Hunan	195	10	584086	9690	7.11	1.10
广 东 Guangdong	55		331924	4521		
广 西 Guangxi	62	8	408372	4496	1.16	24.76
海 南 Hainan	9		60912	829		
重 庆 Chongqing	58	11	132214	1262	32.22	7.34
四 川 Sichuan	259	17	655392	7366	323.04	174.22
贵 州 Guizhou	123	11	468730	4578	140.02	11.32
云 南 Yunnan	80	10	481811	4901	148.15	23.55
西 藏 Tibet	18	4	36152	671	45.69	15.00
陕 西 Shaanxi	199	32	284320	3542	42.82	27.64
甘 肃 Gansu	83	18	173855	2594	0.46	
青 海 Qinghai	15		78400	879	1.87	1.87
宁 夏 Ningxia	27		76688	1080		
新 疆 Xinjiang	20	5	319662	3516	1.00	

四、环境卫生数据
Data by Environmental Health

2-11-1　全国历年县城排水和污水处理情况
National County Seat Drainage and Wastewater Treatment in Past Years

年 份 Year	排水管道长度 （万公里） Length of Drainage Pipelines (10,000 km)	污水年排放量 （亿立方米） Annual Quantity of Wastewater Discharged (100 million m³)	污水处理厂 Wastewater Treatment Plant		污 水 年 处理总量 （亿立方米） Annual Treatment Capacity (100 million m³)	污水处理率 （%） Wastewater Treatment Rate (%)
			座数 （座） Number of Wastewater Treatment Plant (unit)	处理能力 （万立方米／日） Treatment Capacity (10,000 m³/day)		
2000	4.00	43.20	54	55	3.26	7.55
2001	4.40	40.14	54	455	3.31	8.24
2002	4.44	43.58	97	310	3.18	11.02
2003	5.32	41.87	93	426	4.14	9.88
2004	6.01	46.33	117	273	5.20	11.23
2005	6.04	47.40	158	357	6.75	14.23
2006	6.86	54.63	204	496	6.00	13.63
2007	7.68	60.10	322	725	14.10	23.38
2008	8.39	62.29	427	961	19.70	31.58
2009	9.63	65.70	664	1412	27.36	41.64
2010	10.89	72.02	1052	2040	43.30	60.12
2011	12.18	79.52	1303	2409	55.99	70.41
2012	13.67	85.28	1416	2623	62.18	75.24
2013	14.88	88.09	1504	2691	69.13	78.47
2014	16.03	90.47	1555	2882	74.29	82.12
2015	16.79	92.65	1599	2999	78.95	85.22
2016	17.19	92.72	1513	3036	81.02	87.38
2017	18.98	95.07	1572	3218	85.77	90.21
2018	19.98	99.43	1598	3367	90.64	91.16
2019	21.34	102.30	1669	3587	95.71	93.55
2020	22.39	103.76	1708	3770	98.62	95.05
2021	23.84	109.31	1765	3979	105.06	96.11
2022	25.17	114.93	1801	4185	111.41	96.94

2-11-2　县城排水和污水处理(2022年)

地区名称 Name of Regions	污　水 排放量 (万立方米) Annual Quantity of Wastewater Discharged (10,000 m³)	排水管道长度 (公里) Length of Drainage Piplines (km)	污水管道 Sewers	雨水管道 Rainwater Drainage Pipeline	雨污合流管道 Combined Drainage Pipeline	建成区 in Built District	污水处理厂			
							座数 (座) Number of Wastewater Treatment Plant (unit)	二级以上 Second Level or Above	处理能力 (万立方米/日) Treatment Capacity (10,000 m³/day)	二级以上 Second Level or Above
全　国　National Total	1149279	251696	119186	94242	38267	224845	1801	1508	4184.7	3587.6
河　北　Hebei	77212	15147	8095	7052		14559	107	95	366.7	332.5
山　西　Shanxi	37653	8844	4750	3145	950	7945	87	78	144.3	126.0
内蒙古　Inner Mongolia	28923	9534	4746	3367	1420	8674	68	57	107.2	94.7
辽　宁　Liaoning	24923	2717	982	909	827	2323	30	14	87.5	39.0
吉　林　Jilin	15218	2454	1179	1094	182	2414	19	19	52.5	52.5
黑龙江　Heilongjiang	24028	4174	1381	1232	1561	4005	51	51	89.4	89.4
江　苏　Jiangsu	47299	10676	4234	5245	1197	9160	31	28	141.0	134.5
浙　江　Zhejiang	59892	12407	7273	4546	588	10743	48	47	191.9	188.9
安　徽　Anhui	78983	19236	8653	9002	1580	16454	68	67	258.6	251.6
福　建　Fujian	44651	8842	4365	3831	646	8169	46	42	151.9	141.4
江　西　Jiangxi	55951	16976	7730	6671	2576	15295	78	47	168.7	101.7
山　东　Shandong	81673	18582	8616	9560	406	16572	86	85	398.0	394.0
河　南　Henan	103420	20034	8649	7028	4357	17979	134	94	468.8	337.0
湖　北　Hubei	38752	6677	3036	2055	1586	6130	42	36	127.5	103.0
湖　南　Hunan	94686	14821	6335	5074	3412	13466	88	60	291.1	210.1
广　东　Guangdong	41482	5320	2166	1423	1730	4613	41	37	120.9	113.7
广　西　Guangxi	40666	9226	4211	2759	2256	8855	67	63	127.8	124.3
海　南　Hainan	6860	1531	779	541	211	1016	12	4	22.6	8.8
重　庆　Chongqing	14426	3269	1845	1246	179	3134	25	20	53.4	37.4
四　川　Sichuan	74808	15240	7195	5690	2354	13843	147	111	235.3	195.0
贵　州　Guizhou	32213	9493	5563	3230	700	7195	121	121	115.3	115.3
云　南　Yunnan	36726	12833	7211	4415	1206	11397	102	95	124.8	112.5
西　藏　Tibet	4146	1531	457	288	786	1197	49	28	11.3	7.4
陕　西　Shaanxi	29660	6408	3013	1878	1517	5564	72	67	118.4	113.7
甘　肃　Gansu	17668	5883	3287	1959	638	5147	66	62	67.7	62.8
青　海　Qinghai	5325	2112	1009	670	432	1963	37	36	24.2	23.7
宁　夏　Ningxia	8341	1903	189	306	1408	1727	14	13	31.5	30.5
新　疆　Xinjiang	23696	5826	2238	26	3562	5306	65	31	86.5	46.3

County Seat Drainage and Wastewater Treatment(2022)

污水处理厂 Wastewater Treatment Plant				其他污水处理设施 Other Wastewater Treatment Facilities		污水处理总量（万立方米）Total Quantity of Wastewater Treated	再生水 Recycled Water			地区名称
处理量（万立方米）Quantity of Wastewater Treated (10,000 m³)	二级以上 Second Level or Above	干污泥产生量（吨）Quantity of Dry Sludge Produced (ton)	干污泥处置量（吨）Quantity of Dry Sludge Treated (ton)	处理能力（万立方米／日）Treatment Capacity (10,000 m³/day)	处理量（万立方米）Quantity of Wastewater Treated (10,000 m³)	(10,000 m³)	生产能力（万立方米／日）Recycled Water Production Capacity (10,000 m³/day)	利用量（万立方米）Annual Quantity of Wastewater Recycled and Reused (10,000 m³)	管道长度（公里）Length of Piplines (km)	Name of Regions
1104987	947261	2052922	2027827	105.0	9105	1114092	1145.5	177856	6740	全　国
76073	68910	130389	130047			76073	211.8	39684	500	河　北
36791	33392	104138	99867	5.2		36791	74.8	8224	321	山　西
28326	24788	91417	91417			28326	75.5	11159	1591	内蒙古
25180	13106	55673	53005	0.1	5	25184	8.2	1137	41	辽　宁
14965	14965	20149	20149			14965	13.2	734	34	吉　林
23214	23214	42564	42564	0.3	0	23214	6.7	196	28	黑龙江
41806	39719	82329	82329			41806	60.0	9369	141	江　苏
58639	57775	81209	81205	7.7	94	58733	51.8	8825	136	浙　江
75706	74679	98662	98630	3.8	603	76309	49.6	11995	192	安　徽
42935	40246	67378	67336	3.0	575	43510	3.1	27		福　建
52506	33082	75258	75078	9.8	747	53253	3.5	435	7	江　西
80161	80161	186326	186208			80161	255.6	39335	404	山　东
101748	72736	237694	234141	7.3	85	101833	75.6	12912	221	河　南
37310	28907	53649	52017	3.0		37310	27.1	2681	7	湖　北
91211	65852	150800	147663	11.8	136	91348	9.9	226	5	湖　南
39658	37198	63866	62072			39658	18.5	1428		广　东
37688	36439	26002	26002	14.1	2266	39954				广　西
6939	2605	18390	16383	5.5	1224	8163	8.1	168	22	海　南
14488	10073	19192	19192			14488	3.8	226	10	重　庆
68241	55229	122865	122601	23.1	3303	71544	46.5	9240	975	四　川
31316	31316	35526	34944	0.1		31316	13.3	1900	29	贵　州
36023	32459	47265	46361	6.0	6	36029	6.9	211	66	云　南
2609	1718	3157	1016	0.5	56	2665	0.1	44		西　藏
28310	27277	100754	99396	1.3		28310	25.9	2317	26	陕　西
17166	15671	46837	46773			17166	23.5	2096	262	甘　肃
4925	4848	9634	9632			4925		540	41	青　海
8246	7948	25273	25273			8246	9.1	1332	138	宁　夏
22809	12950	56527	56527	2.3	4	22813	63.5	11416	1542	新　疆

2-12-1　全国历年县城市容环境卫生情况

年 份 Year	生活垃圾　Domestic Garbage			
	清运量 （万吨） Quantity of Collected and Transported （10,000 ton）	无害化处理场 （厂）座数 （座） Number of Harmless Treatment Plants/Grounds （unit）	无害化 处理能力 （吨／日） Harmless Treatment Capacity （ton/day）	无害化 处理量 （万吨） Quantity of Harmlessly Treated （10,000 ton）
2000	5560	358	18493	782.52
2001	7851	489	29300	1551.88
2002	6503	460	31582	1056.61
2003	7819	380	29546	1159.35
2004	8182	295	26032	865.13
2005	9535	203	23049	688.78
2006	6266	124	15245	414.30
2007	7110	137	18785	496.56
2008	6794	211	34983	838.69
2009	8085	286	45430	1220.15
2010	6317	448	69310	1732.51
2011	6743	683	103583	2728.72
2012	6838	848	126747	3690.65
2013	6506	992	151615	4298.28
2014	6657	1129	168131	4766.44
2015	6655	1187	181429	5259.94
2016	6666	1273	190672	5680.47
2017	6747	1300	205417	6139.95
2018	6660	1324	220696	6212.38
2019	6871	1378	246729	6609.78
2020	6810	1428	358319	6691.32
2021	6791	1441	338012	6687.44
2022	6705	1343	333306	6653.37

注：自2006年起，生活垃圾填埋场的统计采用新的认定标准，生活垃圾无害化处理数据与往年不可比。

National County Seat Environmental Sanitation in Past Years

粪 便 清运量 (万吨) Volume of Soil Collected and Transported (10,000 ton)	公共厕所 (座) Number of Latrine (unit)	市容环卫专用 车辆设备总数 (辆) Number of Vehicles and Equipment Designated for Municipal Environmental Sanitation (unit)	每 万 人 拥有公厕 (座) Number of Latrine per 10,000 Population (unit)	年 份 Year
1301	31309	13118	2.21	2000
1709	31893	13472	3.54	2001
1659	31282	13817	3.53	2002
1699	33139	15114	3.59	2003
1256	34104	16144	3.54	2004
1312	34753	17697	3.46	2005
710	34563	17367	2.91	2006
2507	36542	19220	2.90	2007
1151	37718	20947	2.90	2008
759	39618	22905	2.96	2009
811	40818	25249	2.94	2010
751	40096	28045	2.80	2011
649	41588	31164	2.09	2012
553	42217	35096	2.77	2013
532	43159	38913	2.76	2014
489	43480	42702	2.78	2015
420	43582	46278	2.82	2016
	45808	54575	2.93	2017
	49059	61436	3.13	2018
	52046	69083	3.28	2019
	55548	74722	3.51	2020
	58674	80064	3.75	2021
	61610	85856	3.95	2022

Note: Since 2006, treatment of domestic garbage through sanitary landfill has adopted new certification standard,so the datas of
      harmless treatmented garbage are not compared with the past years.

2-12-2 县城市容环境卫生(2022年)

地区名称 Name of Regions	道路清扫保洁面积(万平方米) Surface Area of Roads Cleaned and Maintained (10,000 m²)	机械化 Mecha-nization	生活垃圾						
			清运量(万吨) Collected and Trans-ported (10,000 ton)	处理量(万吨) Volume of Treated (10,000 ton)	无害化处理厂(场)数(座) Number of Harmless Treatment Plants/ Grounds (unit)	卫生填埋 Sanitary Landfill	焚烧 Inciner-ation	其他 other	无害化处理能力(吨／日) Harmless Treatment Capacity (ton/day)
全 国 National Total	311052	243884	6704.65	6692.55	1343	957	313	73	333306
河 北 Hebei	22617	20347	398.60	398.60	55	15	37	3	33201
山 西 Shanxi	12370	9696	296.86	294.68	73	67	4	2	10295
内 蒙 古 Inner Mongolia	14968	11601	242.55	242.52	69	68		1	8004
辽 宁 Liaoning	4414	3000	106.35	105.91	22	20		2	3870
吉 林 Jilin	3296	2383	73.45	73.45	17	16	1		3489
黑 龙 江 Heilongjiang	6142	4635	174.91	174.91	40	38	1	1	5168
江 苏 Jiangsu	10425	9215	230.33	230.33	25	4	19	2	15136
浙 江 Zhejiang	10249	7818	226.61	226.61	42	4	25	13	19311
安 徽 Anhui	24437	22093	329.52	329.52	43	14	23	6	19292
福 建 Fujian	8685	5393	260.84	260.84	36	21	12	3	13053
江 西 Jiangxi	20299	17336	368.72	368.72	33	11	20	2	13772
山 东 Shandong	22289	19263	373.42	373.42	59	14	40	5	29870
河 南 Henan	29062	22194	593.71	592.21	73	47	25	1	30404
湖 北 Hubei	8589	6402	194.28	194.28	38	24	8	6	9664
湖 南 Hunan	19038	15005	533.64	533.56	71	52	15	4	19761
广 东 Guangdong	8866	4360	236.02	236.02	36	27	6	3	13876
广 西 Guangxi	9376	4908	244.58	244.58	56	43	12	1	10897
海 南 Hainan	2108	1838	43.74	43.74	2		2		1300
重 庆 Chongqing	2910	2280	83.82	83.82	14	10	3	1	3261
四 川 Sichuan	17400	12804	453.92	453.14	92	64	22	6	18024
贵 州 Guizhou	11214	10746	271.26	268.74	83	59	17	7	13450
云 南 Yunnan	11523	9184	271.70	271.70	79	64	11	4	10887
西 藏 Tibet	1388	239	35.84	35.54	62	62			1259
陕 西 Shaanxi	7906	6302	220.51	219.24	70	63	7		11882
甘 肃 Gansu	6730	5178	180.76	180.69	47	45	2		4881
青 海 Qinghai	2019	1041	62.16	59.36	33	33			1764
宁 夏 Ningxia	3234	2387	43.23	43.23	10	10			1056
新 疆 Xinjiang	9498	6235	153.31	153.20	63	62	1		6482

County Seat Environmental Sanitation(2022)

| Domestic Garbage | | | 无害化处理量 (万吨) Volume of Harmlessly Treated (10,000 ton) | | | | 公共厕所 (座) Number of Latrines (unit) | 三类以上 Grade III and Above | 市容环卫专用车辆设备总数 (辆) Number of Vehicles and Equipment Designated for Municipal Environmental Sanitation (unit) | 地区名称 Name of Regions |
卫生填埋 Sanitary Landfill	焚烧 Inciner-ation	其他 other		卫生填埋 Sanitary Landfill	焚烧 Inciner-ation	其他 other				
127682	194827	10798	6653.37	2822.32	3692.56	138.50	61610	46365	85856	全　国
3555	28646	1000	398.60	38.81	352.20	7.59	5661	4975	7016	河　北
9038	907	350	283.65	228.84	51.74	3.07	2114	1464	4605	山　西
7894		110	242.52	237.23	3.40	1.90	4657	2572	4096	内 蒙 古
3862		8	105.91	95.56	10.26	0.09	1160	457	1256	辽　宁
2989	500		73.45	39.95	33.50		636	457	2177	吉　林
4808	200	160	174.91	124.36	46.44	4.11	1796	486	2669	黑 龙 江
1600	13366	170	230.33		226.07	4.26	2190	2016	2560	江　苏
840	16139	2332	226.61		202.37	24.24	1728	1652	2603	浙　江
2804	15642	846	329.52	20.49	301.52	7.51	3216	2511	3753	安　徽
3897	8596	560	260.84	109.75	145.37	5.72	1717	1379	2112	福　建
2483	11249	40	368.72	40.87	324.68	3.18	3252	3238	6787	江　西
3510	25700	660	373.42	11.98	351.59	9.85	2495	2171	3854	山　东
11944	18420	40	581.47	243.59	337.89		5489	4395	5935	河　南
5144	3840	680	194.28	85.43	95.23	13.63	1201	884	2055	湖　北
11014	8351	396	533.56	266.59	249.47	17.49	2421	1722	3571	湖　南
6878	6500	498	236.02	114.56	111.02	10.44	683	578	2573	广　东
5102	5780	15	244.58	132.84	111.67	0.07	831	705	3002	广　西
	1300		43.74		43.74		366	330	1238	海　南
2301	930	30	83.82	53.74	27.82	2.26	718	644	687	重　庆
6112	11171	741	452.94	148.80	293.18	10.96	3063	2516	4997	四　川
4728	8360	362	268.74	127.40	137.98	3.35	3352	2710	4092	贵　州
5767	3320	1800	255.82	149.22	99.99	6.61	4722	3973	4108	云　南
1259			34.21	34.21			1359	372	741	西　藏
7102	4780		219.24	182.52	36.60	0.12	2697	2017	3255	陕　西
4351	530		180.69	115.24	64.49	0.97	1565	1006	2223	甘　肃
1764			59.36	59.36			710	124	795	青　海
1056			43.23	33.81	8.33	1.09	448	288	563	宁　夏
5882	600		153.20	127.20	26.00		1363	723	2533	新　疆

五、绿色生态数据

Data by Green Ecology

2-13-1　全国历年县城园林绿化情况

计量单位：公顷

年 份 Year	建成区绿化 覆盖面积 Built District Green Coverage Area	建 成 区 绿地面积 Built District Area of Green Space	公园绿地 面　　积 Area of Public Recreational Green Space	公园面积 Park Area
2000	142667	85452	31807	15736
2001	138338	94803	35082	69829
2002	148214	102684	38378	73612
2003	169737	119884	44628	28930
2004	193274	137170	50997	33678
2005	210393	151859	56869	32830
2006	247318	185389	59244	39422
2007	288085	219780	70849	54488
2008	317981	249748	79773	51510
2009	365354	285850	92236	56015
2010	412730	330318	106872	67325
2011	465885	385636	121300	80850
2012	519812	436926	134057	85272
2013	566706	482823	144644	96630
2014	599348	520483	155083	105680
2015	617022	542249	163526	114093
2016	633304	559476	170700	122103
2017	686922	610268	185320	138776
2018	711680	631559	191699	146656
2019	757497	672684	207897	158584
2020	784249	700180	213017	167699
2021	805418	722931	219322	174292
2022	830007	751902	226323	186357

注：1.自2006年起，"公共绿地"统计为"公园绿地"。
　　2.自2006年起，"人均公共绿地面积"统计为以城区人口和城区暂住人口合计为分母计算的"人均公园绿地面积"。

National County Seat Landscaping in Past Years

Measurement Unit: Hectare

人均公园 绿地面积 (平方米) Public Recreational Green Space Per Capita (m²)	建成区绿化 覆盖率 (%) Green Coverage Rate of Built District (%)	建成区 绿地率 (%) Green Space Rate of Built District (%)	年 份 Year
5.71	10.86	6.51	2000
3.88	13.24	9.08	2001
4.32	14.12	9.78	2002
4.83	15.27	10.79	2003
5.29	16.42	11.65	2004
5.67	16.99	12.26	2005
4.98	18.70	14.01	2006
5.63	20.20	15.41	2007
6.12	21.52	16.90	2008
6.89	23.48	18.37	2009
7.70	24.89	19.92	2010
8.46	26.81	22.19	2011
8.99	27.74	23.32	2012
9.47	29.06	24.76	2013
9.91	29.80	25.88	2014
10.47	30.78	27.05	2015
11.05	32.53	28.74	2016
11.86	34.60	30.74	2017
12.21	35.17	31.21	2018
13.10	36.64	32.54	2019
13.44	37.58	33.55	2020
14.01	38.30	34.38	2021
14.50	39.35	35.65	2022

Notes: 1.Since 2006, Public Green Space is changed to Public Recreational Green Space.

2.Since 2006, Public recreational green space per capita has been calculated based on denominator which combines both permanent and temporary residents in urban areas.

2-13-2　县城园林绿化(2022年)

地区名称 Name of Regions		绿化覆盖面积 (公顷) Green Coverage Area (hectare)	建成区 Built District	绿地面积 (公顷) Area of Green Space (hectare)	建成区 Built District
全　国	**National Total**	**1043925**	**830007**	**901298**	**751902**
河　北	Hebei	73467	60550	60781	54888
山　西	Shanxi	33557	28734	28440	25596
内 蒙 古	Inner Mongolia	50376	37992	46354	35417
辽　宁	Liaoning	9435	9168	7926	7657
吉　林	Jilin	15513	9660	13540	8775
黑 龙 江	Heilongjiang	20875	18703	18417	16642
江　苏	Jiangsu	52886	31577	43650	29492
浙　江	Zhejiang	38167	28230	33166	25558
安　徽	Anhui	83904	54726	71989	49953
福　建	Fujian	30460	24665	26875	22679
江　西	Jiangxi	54287	46928	47567	42654
山　东	Shandong	83452	66149	71916	60048
河　南	Henan	81918	73366	69440	63844
湖　北	Hubei	32305	24522	28267	22495
湖　南	Hunan	54884	47992	48604	43934
广　东	Guangdong	33413	24238	26766	22206
广　西	Guangxi	32076	26155	26904	23091
海　南	Hainan	6436	6206	5500	5397
重　庆	Chongqing	13146	8381	11600	7674
四　川	Sichuan	57547	52102	51994	47283
贵　州	Guizhou	51148	34958	46264	33217
云　南	Yunnan	34104	30503	30243	27766
西　藏	Tibet	1177	1075	710	679
陕　西	Shaanxi	29489	23181	24095	20551
甘　肃	Gansu	21041	16442	16846	14551
青　海	Qinghai	6300	5648	5226	4890
宁　夏	Ningxia	9102	7814	8023	7211
新　疆	Xinjiang	33461	30341	30196	27755

County Seat Landscaping(2022)

公园绿地面积（公顷）Area of Public Recreational Green Space (hectare)	公园个数（个）Number of Parks (unit)	门票免费 Free Parks	公园面积（公顷）Park Area (hectare)	地区名称 Name of Regions
226323	**11462**	**10963**	**186357**	**全　国**
14959	1433	1366	13411	河　北
7864	481	441	7392	山　西
10686	490	487	9829	内 蒙 古
2686	83	81	2220	辽　宁
2840	99	98	2451	吉　林
5441	237	221	3899	黑 龙 江
7976	287	287	5381	江　苏
7351	538	536	4871	浙　江
14827	590	576	11951	安　徽
7602	480	457	5709	福　建
13220	854	799	12825	江　西
17407	490	487	13509	山　东
18582	515	485	12773	河　南
6570	415	376	6009	湖　北
13165	539	535	12021	湖　南
7304	374	365	8022	广　东
6784	334	308	5781	广　西
457	63	57	463	海　南
3242	152	147	2542	重　庆
17673	581	562	12461	四　川
10247	387	376	10610	贵　州
7992	908	862	6883	云　南
122	62	54	270	西　藏
5846	337	289	3630	陕　西
5687	247	240	3815	甘　肃
964	51	42	365	青　海
2211	103	103	2074	宁　夏
6617	332	326	5194	新　疆

村镇部分

Statistics for Villages and Small Towns

3-1-1　全国历年建制镇及住宅基本情况

年　份 Year	建 制 镇 统计个数 （万个） Number of Towns （10,000 units）	建成区面积 （万公顷） Surface Area of Build Districts （10,000 hectare）	建 成 区 户籍人口 （亿人） Registered Permanent Population （100 million persons）	非农人口 Nonagriculture Population	建 成 区 暂住人口 （亿人） Temporary Population （100 million persons）	本　年 建设投入 （亿元） Construction Input This Year （100 million RMB）
1990	1.01	82.5	0.61	0.28		156
1991	1.03	87.0	0.66	0.30		192
1992	1.20	97.5	0.72	0.32		284
1993	1.29	111.9	0.79	0.34		458
1994	1.43	118.8	0.87	0.38		616
1995	1.50	138.6	0.93	0.42		721
1996	1.58	143.7	0.99	0.42		915
1997	1.65	155.3	1.04	0.44		821
1998	1.70	163.0	1.09	0.46		872
1999	1.73	167.5	1.16	0.49		980
2000	1.79	182.0	1.23	0.53		1123
2001	1.81	197.2	1.30	0.56		1278
2002	1.84	203.2	1.37	0.60		1520
2003						
2004	1.78	223.6	1.43	0.64		2373
2005	1.77	236.9	1.48	0.66		2644
2006	1.77	312.0	1.40		0.24	3013
2007	1.67	284.3	1.31		0.24	2950
2008	1.70	301.6	1.38		0.25	3285
2009	1.69	313.1	1.38		0.26	3619
2010	1.68	317.9	1.39		0.27	4356
2011	1.71	338.6	1.44		0.26	5018
2012	1.72	371.4	1.48		0.28	5751
2013	1.74	369.0	1.52		0.30	7148
2014	1.77	379.5	1.56		0.31	7172
2015	1.78	390.8	1.60		0.31	6781
2016	1.81	397.0	1.62		0.32	6825
2017	1.81	392.6	1.55			7410
2018	1.83	405.3	1.61			7562
2019	1.87	422.9	1.65			8357
2020	1.88	433.9	1.66			9678
2021	1.91	433.6	1.66			9342
2022	1.92	442.3	1.66			9140

National Summary of Towns and Residential Building in Past Years

住 宅 Residential Building	市政公用设施 Public Facilities	本年住宅竣工建筑面积 (亿平方米) Floor Space Completed of Residential Building This Year (100 million m^2)	年末实有住宅建筑面积 (亿平方米) Total Floor Space of Residential Buildings (year-end) (100 million m^2)	居住人口 (亿人) Resident Population (100 million persons)	人均住宅 建筑面积 (平方米) Per Capita Floor Space (m^2)	年 份 Year
76	15	0.49	12.3	0.61	19.9	1990
84	19	0.54	12.9	0.65	19.8	1991
115	28	0.62	14.8	0.72	20.5	1992
189	56	0.80	15.8	0.78	20.2	1993
265	79	0.90	17.6	0.85	20.6	1994
305	104	1.00	18.9	0.91	20.7	1995
373	116	1.10	20.5	0.97	21.1	1996
382	122	1.06	21.8	1.01	21.5	1997
402	141	1.09	23.3	1.07	21.8	1998
464	160	1.20	24.8	1.13	22.0	1999
530	185	1.41	27.0	1.19	22.6	2000
575	220	1.47	28.6	1.26	22.7	2001
655	265	1.69	30.7	1.32	23.2	2002
						2003
903	437	1.82	33.7	1.40	24.1	2004
1000	476	1.90	36.8	1.43	25.7	2005
1139	580	2.04	39.1	1.40	27.9	2006
1061	614	1.28	38.9		29.7	2007
1211	726	1.33	41.5		30.1	2008
1465	798	1.47	44.2		32.1	2009
1828	1028	1.67	45.1		32.5	2010
2106	1168	1.72	47.3		33.0	2011
2469	1348	1.97	49.6		33.6	2012
3561	1603	2.65	52.0		34.1	2013
3550	1663	2.66	54.0		34.6	2014
3373	1646	2.53	55.4		34.6	2015
3327	1697	2.32	56.7		34.9	2016
3565	1867	3.00	53.9		34.8	2017
3856	1788	3.32	57.9		36.1	2018
4525	1785	2.75	60.4		36.5	2019
5024	2048	2.79	61.4		37.0	2020
4661	1849	2.45	63.2		38.1	2021
4492	1680	2.22	65.2		39.2	2022

3-1-2　全国历年建制镇市政公用设施情况

年　份 Year	年供水总量 （亿立方米） Annual Supply of Water （100 million m³）	生活用水 Domestic Water Consumption	用水人口 （亿人） Population with Access to Water （100 million persons）	供水普及率 （%） Water Coverage Rate （%）	人均日生 活用水量 （升） Per Capita Daily Water Consumption （liter）	道路长度 （万公里） Length of Roads （10,000 km）
1990	24.4	10.0	0.37	60.1	74.3	7.7
1991	29.5	11.6	0.42	63.9	76.1	8.4
1992	35.0	13.6	0.48	65.8	78.1	9.6
1993	39.5	15.8	0.54	68.5	80.7	10.9
1994	47.1	17.7	0.62	71.5	78.3	12.6
1995	53.7	21.5	0.69	74.2	85.5	13.4
1996	62.2	24.7	0.74	75.0	91.7	15.5
1997	68.4	27.0	0.80	76.6	92.6	17.8
1998	72.8	30.0	0.86	79.1	95.1	18.7
1999	81.4	34.3	0.93	80.2	100.8	19.4
2000	87.7	37.1	0.99	80.7	102.7	21.0
2001	91.4	39.6	1.04	80.3	104.0	22.8
2002	97.3	42.3	1.10	80.4	105.4	24.3
2003						
2004	110.7	49.0	1.20	83.6	112.1	27.5
2005	136.5	54.2	1.25	84.7	118.4	30.1
2006	131.0	44.7	1.17	83.8	104.2	26.0
2007	112.0	42.1	1.19	76.6	97.1	21.6
2008	129.0	45.0	1.27	77.8	97.1	23.4
2009	114.6	46.1	1.28	78.3	98.9	24.5
2010	113.5	47.8	1.32	79.6	99.3	25.8
2011	118.6	49.9	1.37	79.8	100.7	27.4
2012	122.2	51.2	1.42	80.8	99.1	29.1
2013	126.2	53.7	1.49	81.7	98.6	31.0
2014	131.6	55.8	1.55	82.8	98.7	32.7
2015	134.8	57.8	1.60	83.8	98.7	34.5
2016	135.3	59.0	1.64	83.9	99.0	35.9
2017	131.9	59.0	1.48	88.1	109.5	33.5
2018	133.7	58.9	1.55	88.1	104.1	37.7
2019	142.6	61.7	1.63	89.0	103.9	40.9
2020	145.2	64.1	1.64	89.1	107.0	43.9
2021	147.1	64.9	1.67	90.3	106.8	45.7
2022	149.5	64.5	1.68	90.8	105.0	47.8

注：1.自2006年起，"公共绿地"统计为"公园绿地"。

　　2.自2006年起，"人均公共绿地面积"统计为以建制镇建成区人口和暂住人口合计为分母计算的"人均公园绿地面积"。

National Municipal Public Facilities of Towns in Past Years

桥梁数 （万座） Number of Bridges (10,000 units)	排水管道 长　度 （万公里） Length of Drainage Piplines (10,000 km)	公园绿地 面　积 （万公顷） Public Green Space (10,000 hectare)	人均公园 绿地面积 （平方米） Public Recreational Green Space Per Capita (m²)	环卫专用 车辆设备 （万辆） Number of Special Vehicles for Environmental Sanitation (10,000 units)	公共厕所 （万座） Number of Latrines (10,000 units)	年　份 Year
2.4	2.7	0.85	1.4	0.5	4.9	1990
2.7	3.2	1.06	1.6	0.7	5.4	1991
3.1	4.0	1.22	1.7	0.8	6.1	1992
3.5	4.8	1.37	1.7	1.1	6.8	1993
3.9	5.5	1.74	2.2	1.2	7.6	1994
4.2	6.2	2.00	2.2	1.6	8.3	1995
4.8	7.5	2.27	2.3	1.7	8.7	1996
5.1	8.1	2.61	2.5	2.0	9.2	1997
5.4	8.8	2.99	2.7	2.3	9.7	1998
5.6	10.0	3.32	2.9	2.5	10.1	1999
6.1	11.1	3.71	3.0	2.9	10.3	2000
6.4	11.9	4.39	3.4	3.2	10.7	2001
6.8	13.0	4.84	3.5	3.3	11.2	2002
						2003
7.2	15.7	6.01	4.2	3.9	11.8	2004
7.7	17.1	6.81	4.6	4.2	12.4	2005
7.2	11.9	3.3	2.4	4.8	9.4	2006
8.3	8.8	2.72	1.8	5.0	9.0	2007
9.1	9.9	3.09	1.9	6.0	12.1	2008
9.9	10.7	3.14	1.9	6.6	11.6	2009
10.0	11.5	3.36	2.0	6.9	9.8	2010
9.7	12.2	3.45	2.0	7.6	10.1	2011
10.4	13.2	3.73	2.1	8.7	10.5	2012
10.8	14.0	4.33	2.4	9.7	14.0	2013
11.0	15.1	4.48	2.4	10.6	11.4	2014
11.4	16.0	4.69	2.5	11.5	11.9	2015
11.1	16.6	4.79	2.5	12.0	11.7	2016
8.4	16.4	5.24	3.1	11.5	12.1	2017
8.3	17.7	4.99	2.8	11.4	11.8	2018
8.4	18.8	4.96	2.7	12.1	12.9	2019
8.1	19.8	5.01	2.7	12.0	13.5	2020
7.4	21.1	4.97	2.7	11.6	12.7	2021
7.2	21.8	4.99	2.7	11.5	12.6	2022

Notes: 1.Since 2006, Public Green Space is changed to Public Recreational Green Space.
2.Since 2006, Public Recreational Green Space Per Capita has been calculated based on denominator which combines both permanent and temporary residents in built area of town.

3-1-3 全国历年乡及住宅基本情况

年 份 Year	乡统计个 数（万个） Number of Townships (10,000 units)	建成区面 积（万公顷） Surface Area of Build Districts (10,000 hectare)	建成区户籍人口（亿人） Registered Permanent Population (100 million persons)	非农人口 Nonagriculture Population	建成区暂住人口（亿人） Temporary Population (100 million persons)	本年建设投入（亿元） Construction Input This Year (100 million RMB)
1990	4.02	110.1	0.72	0.17		121
1991	3.90	109.3	0.70	0.16		136
1992	3.72	98.1	0.66	0.15		168
1993	3.64	99.9	0.65	0.15		191
1994	3.39	101.2	0.62	0.14		234
1995	3.42	103.7	0.63	0.15		260
1996	3.15	95.2	0.60	0.14		296
1997	3.03	95.7	0.60	0.14		296
1998	2.91	93.7	0.59	0.15		316
1999	2.87	92.6	0.59	0.15		325
2000	2.76	90.7	0.58	0.14		300
2001	2.35	79.7	0.53	0.14		283
2002	2.26	79.1	0.52	0.14		325
2003						
2004	2.18	78.1	0.53	0.15		344
2005	2.07	77.8	0.52	0.14		377
2006	1.46	92.83	0.35		0.03	355
2007	1.42	75.89	0.34		0.03	352
2008	1.41	81.15	0.34		0.03	438
2009	1.39	75.76	0.33		0.03	471
2010	1.37	75.12	0.32		0.03	558
2011	1.29	74.19	0.31		0.02	535
2012	1.27	79.55	0.31		0.02	634
2013	1.23	73.69	0.31		0.02	706
2014	1.19	72.23	0.30		0.02	671
2015	1.15	70.00	0.29		0.02	559
2016	1.09	67.30	0.28		0.02	524
2017	1.03	63.38	0.25			653
2018	1.02	65.39	0.25			621
2019	0.95	62.95	0.24			665
2020	0.89	61.70	0.24			780
2021	0.82	58.78	0.22			596
2022	0.80	56.85	0.21			481

注：2006年以后，本表统计范围由原来的集镇变为乡。

National Summary of Townships and Residential Building in Past Years

住 宅 Residential Building	市政公 用设施 Public Facilities	本年住宅竣工 建筑面积 （亿平方米） Floor Space Completed of Residential Building This Year (100 million m²)	年末实有住 宅建筑面积 （亿平方米） Total Floor Space of Residential Buildings (year-end) (100 million m²)	居住人口 （亿人） Resident Population (100 million persons)	人均住宅 建筑面积 （平方米） Per Capita Floor Space (m²)	年 份 Year
61	7	0.52	13.8	0.72	19.1	1990
67	8	0.53	13.8	0.70	19.8	1991
76	10	0.55	13.4	0.65	20.6	1992
85	13	0.49	13.3	0.64	20.6	1993
113	16	0.51	12.8	0.63	20.3	1994
133	22	0.57	12.7	0.62	20.5	1995
151	26	0.59	12.2	0.58	21.0	1996
155	33	0.56	12.3	0.59	21.0	1997
175	37	0.57	12.3	0.58	21.4	1998
193	36	0.66	12.8	0.58	22.1	1999
175	35	0.60	12.6	0.56	22.6	2000
167	33	0.55	12.0	0.52	23.0	2001
188	39	0.57	12.0	0.51	23.6	2002
						2003
188	48	0.56	12.5	0.50	24.9	2004
186	55	0.56	12.8	0.50	25.5	2005
145	66	0.40	9.1	0.35	25.9	2006
147	75	0.26	9.1		27.1	2007
187	99	0.28	9.2		27.2	2008
212	101	0.29	9.4		28.8	2009
262	129	0.35	9.7		29.9	2010
267	122	0.32	9.5		30.3	2011
306	152	0.36	9.6		30.5	2012
365	153	0.39	9.6		31.2	2013
332	132	0.36	9.3		31.2	2014
285	134	0.32	9.0		31.2	2015
260	136	0.29	8.7		31.2	2016
319	175	0.56	7.9		31.5	2017
304	175	0.39	8.4		33.2	2018
300	178	0.44	8.3		33.9	2019
364	171	0.44	8.4		35.4	2020
285	151	0.33	8.1		37.0	2021
203	144	0.18	7.7		36.5	2022

Note: Since 2006, coverage of the statistics in the table is changed to township.

3-1-4　全国历年乡市政公用设施情况

年 份 Year	年供水总量 （亿立方米） Annual Supply of Water (100 million m³)	生活用水 Domestic Water Consumption	用水人口 （亿人） Population with Access to Water (100 million persons)	供水普及率 （%） Water Coverage Rate (%)	人均日生 活用水量 （升） Per Capita Daily Water Consumption (liter)	道路长度 （万公里） Length of Roads (10,000 km)
1990	10.8	5.0	0.26	35.7	53.4	15.2
1991	12.3	5.1	0.27	39.3	51.1	14.9
1992	12.7	5.2	0.28	42.6	50.6	14.2
1993	12.3	5.6	0.27	40.6	58.2	14.2
1994	12.8	6.0	0.30	47.8	55.5	14.0
1995	13.7	6.4	0.32	49.9	55.4	14.4
1996	13.9	6.6	0.29	49.0	61.1	14.4
1997	16.2	7.3	0.31	52.3	63.7	14.5
1998	17.2	7.9	0.33	55.5	66.4	14.3
1999	17.3	8.8	0.35	58.2	69.7	14.4
2000	16.8	8.8	0.35	60.1	69.2	13.7
2001	15.7	8.2	0.32	61.0	69.3	12.1
2002	16.4	8.4	0.32	62.1	71.9	12.1
2003						
2004	17.4	9.5	0.35	65.8	74.8	12.6
2005	17.5	9.6	0.35	67.2	75.6	12.4
2006	25.8	6.3	0.22	63.4	78.0	7.0
2007	11.9	6.0	0.21	59.1	76.1	6.2
2008	11.9	6.3	0.23	62.6	75.5	6.4
2009	11.4	6.5	0.22	63.5	79.5	6.3
2010	11.8	6.8	0.23	65.6	81.4	6.6
2011	11.5	6.7	0.22	65.7	82.4	6.5
2012	12.0	6.9	0.22	66.7	83.9	6.7
2013	11.5	6.8	0.22	68.2	82.8	6.8
2014	11.3	6.7	0.22	69.3	83.1	7.0
2015	11.2	6.7	0.22	70.4	84.3	7.1
2016	11.2	6.7	0.22	71.9	85.3	7.3
2017	12.6	7.2	0.19	78.8	104.3	6.6
2018	12.1	6.6	0.20	79.2	91.9	8.1
2019	12.9	6.6	0.20	80.5	93.3	8.7
2020	13.5	6.8	0.19	83.9	97.0	8.9
2021	13.2	6.5	0.18	84.2	98.7	8.8
2022	12.8	6.3	0.17	84.7	99.5	8.9

注：1.自2006年起，"公共绿地"统计为"公园绿地"。
　　2.自2006年起，"人均公共绿地面积"统计为以乡建成区人口和暂住人口合计为分母计算的"人均公园绿地面积"。

National Municipal Public Facilities of Townships in Past Years

桥梁数 （万座） Number of Bridges (10,000 units)	排水管道 长 度 （万公里） Length of Drainage Piplines (10,000 km)	公园绿 地面积 （万公顷） Public Green Space (10,000 hectare)	人均公园 绿地面积 （平方米） Public Recreational Green Space Per Capita (m²)	环卫专用 车辆设备 （万辆） Number of Special Vehicles for Environmental Sanitation (10,000 units)	公共厕所 （万座） Number of Latrines (10,000 units)	年 份 Year
3.3	2.3	0.64	0.88	0.16	5.33	1990
3.5	2.3	0.83	1.19	0.26	5.74	1991
3.7	2.5	0.87	1.32	0.31	5.89	1992
3.4	2.4	0.91	1.4	0.29	5.77	1993
3.3	2.6	1.11	1.76	0.42	6.18	1994
3.4	3.7	1.09	1.73	0.50	6.42	1995
3.4	3.2	1.08	1.79	0.50	6.12	1996
3.5	3.1	1.15	1.91	0.54	6.25	1997
3.5	3.2	1.31	2.22	0.62	6.21	1998
3.5	3.2	1.32	2.23	0.65	6.04	1999
3.4	3.3	1.35	2.33	0.68	5.86	2000
2.9	3.1	1.36	2.56	0.70	5.03	2001
2.9	3.6	1.31	2.54	0.75	5.05	2002
						2003
2.8	4.3	1.41	2.57	0.77	4.58	2004
2.9	4.3	1.37	2.65	0.80	4.57	2005
2.2	1.9	0.29	0.85	0.88	2.92	2006
2.7	1.1	0.24	0.66	1.04	2.76	2007
2.6	1.2	0.26	0.72	1.30	3.34	2008
2.8	1.4	0.30	0.84	1.34	2.96	2009
2.7	1.4	0.31	0.88	1.45	2.75	2010
2.6	1.4	0.30	0.90	1.53	2.58	2011
2.6	1.5	0.32	0.95	1.85	3.08	2012
2.6	1.6	0.35	1.08	2.56	3.94	2013
2.7	1.6	0.34	1.07	2.37	3.19	2014
2.7	1.7	0.34	1.10	2.41	3.04	2015
2.6	1.8	0.33	1.11	2.50	2.99	2016
1.9	1.9	0.40	1.65	2.76	3.18	2017
1.9	2.4	0.37	1.50	2.80	3.55	2018
1.8	2.5	0.38	1.59	2.93	3.91	2019
1.7	2.4	0.40	1.76	2.86	3.88	2020
1.6	2.3	0.36	1.69	2.77	3.65	2021
1.5	2.3	0.37	1.82	2.70	3.56	2022

Notes: 1.Since 2006, Public Green Space is changed to Public Recreational Green Space.
    2.Since 2006, Public recreational green space per capita has been calculated based on denominator which combines both permanent and temporary residents in build area of township.

3-1-5 全国历年村庄基本情况

年 份 Year	村庄统计 个 数 （万个） Number of Villages (10,000unit)	村庄现状 用地面积 （万公顷） Area of Villages (10,000 hectare)	村庄户籍 人 口 （亿人） Registered Permanent Population (100 million persons)	非农人口 Non- agriculture Population	村庄暂住 人 口 （亿人） Temporary Population (100 million persons)	本年建设 投 入 （亿元） Construction Input This Year (100 million RMB)	住 宅 Residential Building
1990	377.3	1140.1	7.92	0.16		662	545
1991	376.2	1127.2	8.00	0.16		744	618
1992	375.5	1187.7	8.06	0.16		793	624
1993	372.1	1202.7	8.13	0.17		906	659
1994	371.3	1243.8	8.15	0.18		1175	885
1995	369.5	1277.1	8.29	0.20		1433	1089
1996	367.6	1336.1	8.18	0.19		1516	1176
1997	365.9	1366.4	8.18	0.20		1538	1175
1998	355.8	1372.6	8.15	0.21		1585	1220
1999	359.0	1346.3	8.13	0.22		1607	1245
2000	353.7	1355.3	8.12	0.24		1572	1203
2001	345.9	1396.1	8.06	0.25		1558	1145
2002	339.6	1388.8	8.08	0.26		2002	1288
2003							
2004	320.7	1362.7	7.95	0.32		2064	1243
2005	313.7	1404.2	7.87	0.31		2304	1374
2006	270.9		7.14		0.23	2723	1524
2007	264.7	1389.9	7.63		0.28	3544	1923
2008	266.6	1311.7	7.72		0.31	4294	2558
2009	271.4	1362.8	7.70		0.28	5400	3456
2010	273.0	1399.2	7.69		0.29	5692	3412
2011	266.9	1373.8	7.64		0.28	6204	3773
2012	267.0	1409.0	7.63		0.28	7420	4312
2013	265.0	1394.3	7.62		0.28	8183	4898
2014	270.2	1394.1	7.63		0.28	8088	5020
2015	264.5	1401.3	7.65		0.28	8203	5059
2016	261.7	1392.2	7.63		0.27	8321	5045
2017	244.9		7.56			9168	5271
2018	245.2		7.71			9830	5355
2019	251.3		7.76			10167	5529
2020	236.3		7.77			11503	5670
2021	236.1		7.72			10255	5142
2022	233.2		7.72			8849	4438

National Summary of Villages in Past Years

市政公用设施 Public Facilities	本年住宅竣工建筑面积 (亿平方米) Floor Space Completed of Residential Building This Year (100 million m²)	年末实有住宅建筑面积 (亿平方米) Total Floor Space of Residential Buildings (year-end) (100 million m²)	居住人口 (亿人) Resident Population (100 million persons)	人均住宅建筑面积 (平方米) Per Capita Floor Space (m²)	道路长度 (万公里) Length of Roads (10,000 km)	桥梁数 (万座) Number of Bridges (10,000 units)	年 份 Year
33	4.82	159.3	7.84	20.3	262.1		1990
26	5.54	163.3	7.95	20.5	240.0	37.7	1991
32	4.86	167.4	8.01	20.9	262.9	40.2	1992
57	4.38	170.0	8.12	20.9	268.7	43.0	1993
65	4.49	169.1	7.90	21.4	263.2	43.0	1994
104	4.95	177.7	8.06	22.0	275.0	44.7	1995
106	4.96	182.4	8.13	22.4	279.3	44.1	1996
136	4.66	185.9	8.12	22.9	283.2	44.7	1997
139	4.73	189.2	8.07	23.5	290.3	43.4	1998
152	4.62	192.8	8.04	24.0	287.3	45.7	1999
139	4.47	195.2	8.02	24.3	287.0	46.3	2000
160	4.28	199.1	7.97	25.0	283.6	46.7	2001
368	4.39	202.5	7.94	25.5	287.3	47.1	2002
							2003
342	4.22	205.0	7.75	26.5	285.1	57.8	2004
380	4.42	208.0	7.72	26.9	304.0	58.0	2005
501	4.75	202.9	7.14	28.4	221.9	50.7	2006
616	3.65	222.7		29.2			2007
793	4.10	227.2		29.4			2008
863	4.91	237.0		30.8			2009
1105	4.56	242.6		31.6			2010
1216	4.86	245.1		32.1			2011
1660	5.25	247.8		32.5			2012
1850	5.46	250.6		32.9	228.0		2013
1707	5.46	253.4		33.2	234.1		2014
1919	5.66	255.2		33.4	239.3		2015
2120	5.32	256.1		33.6	246.3		2016
2529	9.65	246.2		32.6	285.3		2017
3053	7.81	252.2		32.7	304.8		2018
3100	7.12	255.3		32.9	320.6		2019
3590	7.56	266.5		34.3	335.8		2020
3357	5.47	267.3		34.6	348.7		2021
2660	4.38	269.8		34.9	357.4		2022

3-2-1 建制镇市政公用设施水平(2022年)

地区名称 Name of Regions	人口密度 (人/平方公里) Population Density (person/km²)	人均日生 活用水量 (升) Per Capita Daily Water Consumption (liter)	供水 普及率 (%) Water Coverage Rate (%)	燃气 普及率 (%) Gas Coverage Rate (%)	人均道 路面积 (平方米) Road Surface Area Per Capita (m²)	排水管 道暗渠 密度 (公里/平方公里) Density of Drains (km/km²)
全 国 National Total	**4187**	**105.05**	**90.76**	**59.16**	**17.21**	**7.66**
北 京 Beijing	4087	130.84	92.90	73.70	13.28	6.06
天 津 Tianjin	3535	91.16	96.24	89.07	14.85	5.59
河 北 Hebei	3836	89.63	93.13	76.93	12.46	4.55
山 西 Shanxi	3732	82.91	82.10	35.30	14.84	6.31
内 蒙 古 Inner Mongolia	2007	88.14	82.94	28.16	20.24	3.16
辽 宁 Liaoning	3150	121.78	81.25	30.93	16.55	4.32
吉 林 Jilin	2520	96.80	93.22	50.98	17.77	4.12
黑 龙 江 Heilongjiang	2814	87.77	86.68	17.21	17.91	4.14
上 海 Shanghai	6105	122.44	93.25	76.74	8.21	4.86
江 苏 Jiangsu	5144	104.91	98.71	94.36	20.32	11.44
浙 江 Zhejiang	4645	120.86	91.90	57.14	17.78	9.72
安 徽 Anhui	3984	102.85	86.52	54.39	20.54	8.50
福 建 Fujian	4671	116.51	93.23	70.62	17.15	8.34
江 西 Jiangxi	4041	97.45	85.00	45.34	18.29	9.08
山 东 Shandong	3921	83.82	93.35	73.33	17.08	7.59
河 南 Henan	4357	99.24	86.95	37.48	18.01	7.44
湖 北 Hubei	3727	101.50	89.51	49.67	20.03	8.38
湖 南 Hunan	4231	109.14	82.78	43.63	14.27	7.34
广 东 Guangdong	4637	136.80	94.00	79.46	16.60	9.12
广 西 Guangxi	5115	110.11	91.57	78.35	20.17	9.26
海 南 Hainan	4350	95.38	86.12	78.42	17.27	8.24
重 庆 Chongqing	5284	91.36	96.60	75.96	9.55	8.66
四 川 Sichuan	4512	101.85	89.59	70.09	17.74	8.09
贵 州 Guizhou	3653	98.92	90.05	14.75	20.94	7.90
云 南 Yunnan	4831	97.84	95.27	13.53	16.59	8.56
西 藏 Tibet	3698	940.19	60.65	23.75	31.57	4.79
陕 西 Shaanxi	4227	83.81	90.05	30.61	17.33	7.52
甘 肃 Gansu	3585	79.81	90.76	14.48	18.44	5.90
青 海 Qinghai	3917	87.18	92.58	29.71	21.61	6.05
宁 夏 Ningxia	2982	95.69	97.54	52.54	17.49	6.67
新 疆 Xinjiang	3063	98.86	90.52	24.09	32.99	5.07
新疆生产 建设兵团 Xinjiang Production and Construction Corps	3105	131.23	97.32	58.55	20.65	4.96

Level of Municipal Public Facilities of Built-up Area of Towns (2022)

污 水 处理率 (%) Wastewater Treatment Rate (%)	污水处理 厂集中处 理 率 (%) Centralized Treatment Rate of Wastewater Treatment Plants (%)	人均公 园绿地 面 积 (平方米) Public Recreational Green Space Per Capita (m²)	绿 化 覆盖率 (%) Green Coverage Rate (%)	绿地率 (%) Green Space Rate (%)	生活垃圾 处理率 (%) Domestic Garbage Treatment Rate (%)	无害化 处理率 (%) Domestic Garbage Harmless Treatment Rate	地区名称 Name of Regions
64.86	55.54	2.69	16.97	10.93	92.34	80.38	全　国
69.57	58.13	2.71	24.97	17.35	87.74	81.68	北　京
78.18	74.27	2.25	15.23	9.30	99.59	99.59	天　津
63.42	44.43	1.55	14.82	8.78	99.32	98.50	河　北
32.32	23.76	1.56	15.86	8.40	57.87	18.41	山　西
43.82	40.79	3.69	13.77	8.80	51.34	41.13	内 蒙 古
38.86	35.27	1.11	14.81	7.66	71.24	41.21	辽　宁
52.39	50.70	2.42	11.62	7.03	99.07	92.31	吉　林
33.68	26.86	1.53	9.16	6.39	61.12	52.77	黑 龙 江
64.53	58.46	2.95	18.05	11.35	97.67	92.51	上　海
88.20	82.90	6.98	30.25	24.64	99.06	96.37	江　苏
73.84	60.39	2.57	19.00	13.00	99.11	98.17	浙　江
63.23	53.70	1.82	18.46	10.51	99.54	98.68	安　徽
78.69	62.19	4.66	22.42	16.33	100.00	100.00	福　建
46.20	35.87	1.81	14.42	9.80	92.48	71.10	江　西
72.89	56.49	4.67	24.08	15.67	99.09	96.04	山　东
37.34	33.67	1.71	17.39	6.97	80.80	57.24	河　南
71.03	56.63	1.80	15.55	8.36	96.97	90.87	湖　北
55.99	40.91	2.35	20.12	13.38	87.83	58.44	湖　南
70.58	66.18	3.02	13.81	9.59	97.32	91.34	广　东
59.34	48.17	2.77	16.61	10.81	96.92	72.19	广　西
66.38	46.05	1.28	17.39	11.98	97.70	90.03	海　南
87.30	79.60	0.67	12.23	7.38	96.71	76.17	重　庆
70.66	62.88	1.02	7.37	5.28	96.75	84.67	四　川
61.16	51.40	1.33	13.01	7.74	90.45	57.97	贵　州
22.36	18.16	0.78	8.77	5.79	80.85	48.22	云　南
2.37	2.10	0.34	8.87	1.97	80.00	32.99	西　藏
47.39	39.60	1.42	8.81	5.95	74.07	37.15	陕　西
29.15	23.14	1.21	13.22	8.07	66.00	52.49	甘　肃
23.16	21.90	0.20	11.44	6.97	72.58	37.16	青　海
77.01	72.01	2.22	15.21	9.42	99.13	88.60	宁　夏
31.35	19.48	3.73	17.93	13.78	86.95	56.37	新　疆
61.41	61.29	2.69	21.54	15.89	69.87	50.93	新疆生产 建设兵团

3-2-2　建制镇基本情况(2022年)

地区名称 Name of Regions	建制镇 个 数 (个) Number of Towns (unit)	建成区 面 积 (公顷) Surface Area of Built- up Districts (hectare)	建 成 区 户籍人口 (万人) Registered Permanent Population (10,000 persons)	建 成 区 常住人口 (万人) Permanant Population (10,000 persons)
全　国　**National Total**	19245	4422970.51	16629.05	18520.65
北　京　Beijing	113	29390.85	68.38	120.12
天　津　Tianjin	113	45264.32	119.33	160.02
河　北　Hebei	1129	178676.74	645.66	685.38
山　西　Shanxi	531	86609.87	293.03	323.20
内 蒙 古　Inner Mongolia	436	115229.85	219.00	231.30
辽　宁　Liaoning	612	96287.05	286.56	303.29
吉　林　Jilin	390	79021.71	221.54	199.17
黑 龙 江　Heilongjiang	486	88702.08	274.99	249.64
上　海　Shanghai	101	141604.18	363.31	864.48
江　苏　Jiangsu	656	271924.33	1219.05	1398.84
浙　江　Zhejiang	576	218212.83	701.61	1013.59
安　徽　Anhui	914	264074.30	1033.73	1052.19
福　建　Fujian	562	147042.47	620.45	686.80
江　西　Jiangxi	732	148361.39	581.46	599.47
山　东　Shandong	1056	399793.92	1449.85	1567.63
河　南　Henan	1088	297777.94	1288.12	1297.41
湖　北　Hubei	702	232163.51	846.50	865.29
湖　南　Hunan	1070	255937.59	1086.78	1082.94
广　东　Guangdong	1003	354418.08	1268.78	1643.43
广　西　Guangxi	702	100698.39	548.15	515.12
海　南　Hainan	156	27576.33	107.02	119.95
重　庆　Chongqing	589	79696.54	413.64	421.10
四　川　Sichuan	1848	250173.50	1021.74	1128.78
贵　州　Guizhou	772	147891.93	562.60	540.20
云　南　Yunnan	588	78796.11	368.58	380.65
西　藏　Tibet	74	3455.92	8.39	12.78
陕　西　Shaanxi	923	124672.85	526.94	526.98
甘　肃　Gansu	792	68633.42	240.93	246.05
青　海　Qinghai	104	8899.08	35.69	34.86
宁　夏　Ningxia	77	21650.59	52.94	64.56
新　疆　Xinjiang	313	45574.66	126.25	139.60
新疆生产 建设兵团　Xinjiang Production and Construction Corps	37	14758.18	28.04	45.83

Summary of Towns(2022)

| 规划建设管理 Planning and Administer | | | | | | 地区名称 |
设有村镇建设管理机构的个数（个）Number of Towns with Construction Management Institution (unit)	村镇建设管理人员（人）Number of Construction Management Personnel (person)	专职人员 Full-time Staff	有总体规划的建制镇个数（个）Number of Towns with Master Plans (unit)	本年编制 Compiled This Year	本年规划编制投入（万元）Input in Planning this Year (10,000 RMB)	Name of Regions
17811	94448	59362	16887	910	310075.81	全　国
108	1178	633	96	1	5483.47	北　京
111	716	414	62	5	1631.74	天　津
990	3248	1953	839	45	2723.77	河　北
421	1083	592	372	8	2091.30	山　西
394	1382	892	392	15	1975.00	内　蒙　古
604	1634	1090	525	18	1336.91	辽　宁
381	1152	836	273	13	7819.60	吉　林
466	1121	647	386	26	595.50	黑　龙　江
101	1200	766	85	4	2735.87	上　海
653	7292	4849	635	43	15233.55	江　苏
553	5355	3386	545	40	23127.16	浙　江
848	4423	2749	854	59	26520.62	安　徽
526	2131	1442	524	10	20035.48	福　建
705	2861	1819	700	46	12024.34	江　西
1056	7573	4943	989	73	8407.40	山　东
1071	7335	4762	935	41	10939.13	河　南
687	4719	2969	664	29	13621.63	湖　北
975	5535	3327	980	63	25108.42	湖　南
961	9631	5573	910	53	19633.78	广　东
697	4370	2660	670	15	6421.71	广　西
149	654	482	152		8074.33	海　南
589	2685	1862	555	50	1396.70	重　庆
1528	5171	3430	1425	55	44931.49	四　川
762	2740	1954	712	73	5537.90	贵　州
580	2717	1641	556	18	15982.32	云　南
12	36	9	47	4	307.00	西　藏
818	2374	1390	810	60	9857.55	陕　西
619	2312	1296	708	26	6261.04	甘　肃
78	111	67	99	5	388.00	青　海
70	308	148	69	4	8080.50	宁　夏
264	1006	517	288	6	1554.60	新　疆
34	395	264	30	2	238.00	新疆生产建设兵团

3-2-3 建制镇供水(2022年)

地区名称 Name of Regions	集中供水的 建制镇个数 (个) Number of Towns with Access to Piped Water (unit)	占全部建制镇的比例 (%) Percentage of Total Rate (%)	公共供水综合生产能力 (万立方米/日) Integrated Production Capacity of Public Water Supply Facilities (10,000 m³/day)	自备水源单位综合生产能力 (万立方米/日) Integrated Production Capacity of Self-built Water Supply Facilities (10,000 m³/day)	年供水总量 (万立方米) Annual Supply of Water (10,000 m³)
全　国　National Total	18998	98.72	11181.28	2601.06	1494967.91
北　京　Beijing	113	100.00	70.95	58.63	11254.00
天　津　Tianjin	110	97.35	79.10	10.23	11641.05
河　北　Hebei	1098	97.25	198.99	132.42	46796.27
山　西　Shanxi	521	98.12	75.06	24.36	19188.29
内　蒙　古　Inner Mongolia	430	98.62	65.42	22.23	11787.04
辽　宁　Liaoning	531	86.76	128.08	40.47	20172.77
吉　林　Jilin	388	99.49	88.63	29.90	12355.58
黑　龙　江　Heilongjiang	477	98.15	52.76	24.74	9735.28
上　海　Shanghai	101	100.00	850.23	80.76	107305.31
江　苏　Jiangsu	656	100.00	1185.76	107.17	127569.77
浙　江　Zhejiang	572	99.31	815.07	159.37	125389.32
安　徽　Anhui	914	100.00	514.11	117.77	72607.80
福　建　Fujian	561	99.82	365.62	79.71	72409.31
江　西　Jiangxi	728	99.45	325.64	50.06	37623.21
山　东　Shandong	1056	100.00	635.76	266.17	123822.22
河　南　Henan	1086	99.82	837.17	258.18	94707.89
湖　北　Hubei	702	100.00	653.93	176.47	61993.17
湖　南　Hunan	1052	98.32	929.44	147.53	70503.40
广　东　Guangdong	996	99.30	1541.05	368.86	212481.04
广　西　Guangxi	702	100.00	184.05	58.53	32614.67
海　南　Hainan	156	100.00	26.96	4.96	5484.01
重　庆　Chongqing	589	100.00	132.13	21.09	22579.64
四　川　Sichuan	1840	99.57	637.63	148.50	64439.69
贵　州　Guizhou	766	99.22	253.37	58.16	30736.41
云　南　Yunnan	588	100.00	151.69	60.96	26087.08
西　藏　Tibet	49	66.22	11.99	1.37	3701.58
陕　西　Shaanxi	920	99.67	88.73	38.64	24842.49
甘　肃　Gansu	776	97.98	116.63	18.47	13859.45
青　海　Qinghai	95	91.35	38.05	4.03	2252.51
宁　夏　Ningxia	77	100.00	21.90	5.42	3457.90
新　疆　Xinjiang	312	99.68	91.64	23.01	11868.24
新疆生产建设兵团　Xinjiang Production and Construction Corps	36	97.30	13.73	2.92	3701.52

Water Supply of Towns(2022)

年生活 用水量 Annual Domestic Water Consumption	年生产 用水量 Annual Water Consumption for Production	供水管道 长 度 （公里） Length of Water Supply Pipelines (km)	本年新增 Added This Year	用水人口 （万人） Population with Access to Water (10,000 persons)	地区名称 Name of Regions
644500.76	705437.54	677625.48	25014.33	16809.35	全　国
5329.50	4075.84	4181.69	43.00	111.60	北　京
5123.85	6317.92	4014.96	4.96	153.99	天　津
20882.00	21830.32	14668.06	324.31	638.32	河　北
8029.60	9014.51	14700.54	452.88	265.33	山　西
6171.77	4161.31	7890.78	151.89	191.84	内 蒙 古
10953.20	8091.08	8523.40	244.44	246.42	辽　宁
6559.83	5059.84	8030.94	70.89	185.67	吉　林
6932.48	2328.70	8150.84	34.70	216.39	黑 龙 江
36024.74	56690.85	8783.62	94.72	806.12	上　海
52869.33	69602.57	50023.05	903.00	1380.73	江　苏
41091.28	73996.71	35820.69	1025.10	931.51	浙　江
34172.08	28248.75	43760.36	2208.05	910.32	安　徽
27228.57	39023.42	18579.84	774.33	640.30	福　建
18124.11	15840.12	20425.44	1149.68	509.53	江　西
44772.29	72254.20	37765.19	1378.38	1463.39	山　东
40862.78	43947.51	50386.08	1919.94	1128.07	河　南
28696.28	25761.95	45951.72	1884.05	774.55	湖　北
35710.37	27576.12	45949.21	1473.39	896.43	湖　南
77133.57	102573.87	62864.52	2469.17	1544.83	广　东
18957.95	10497.30	14284.38	490.36	471.69	广　西
3596.13	1375.56	5874.63	197.21	103.30	海　南
13563.92	7800.12	9394.05	436.11	406.77	重　庆
37593.61	22533.61	52792.25	2441.91	1011.29	四　川
17565.09	10106.45	31956.51	2033.11	486.48	贵　州
12950.59	11225.83	22874.19	1077.60	362.65	云　南
2659.67	981.09	3345.72	59.40	7.75	西　藏
14517.88	9289.56	17860.65	848.14	474.56	陕　西
6505.69	5363.54	10349.07	280.42	223.33	甘　肃
1026.85	1159.14	1708.71	8.11	32.27	青　海
2199.60	944.51	2936.33	110.42	62.98	宁　夏
4559.99	6678.69	12254.12	303.80	126.37	新　疆
2136.16	1086.55	1523.94	120.86	44.60	新疆生产 建设兵团

3-2-4　建制镇燃气、供热、道路桥梁(2022年)

地区名称 Name of Regions	用气人口 （万人） Population with Access to Gas (10,000 persons)	集中供热面积 （万平方米） Area of Centrally Heated District (10,000 m²)	道路长度 （公里） Length of Roads (km)	本年新增 Added This Year	本年更新改造 Renewal and Upgrading during the Reported Year
全　国　National Total	10956.06	49952.53	477679.79	18506.74	21613.09
北　京　Beijing	88.53	2120.19	2583.92	41.70	69.31
天　津　Tianjin	142.52	4790.11	3362.01	7.41	70.53
河　北　Hebei	527.29	3768.38	15289.23	188.69	686.88
山　西　Shanxi	114.08	2757.56	7990.75	155.55	312.96
内　蒙　古　Inner Mongolia	65.13	3699.09	7337.94	154.97	113.58
辽　宁　Liaoning	93.82	3669.44	8558.99	234.82	450.88
吉　林　Jilin	101.54	3818.15	5819.84	124.37	204.77
黑　龙　江　Heilongjiang	42.97	2585.16	7861.27	9.10	138.58
上　海　Shanghai	663.39	171.86	6188.52	52.49	136.66
江　苏　Jiangsu	1320.01	89.22	39209.34	399.83	1348.89
浙　江　Zhejiang	579.17	379.93	22977.79	551.98	1151.03
安　徽　Anhui	572.33		32358.84	1575.44	1116.56
福　建　Fujian	485.00		17467.00	575.81	612.81
江　西　Jiangxi	271.81	125.54	17143.71	985.61	1291.06
山　东　Shandong	1149.55	13340.55	37597.18	1448.35	2527.65
河　南　Henan	486.24	1410.90	37764.83	2182.86	1714.76
湖　北　Hubei	429.81	385.16	27686.51	1500.14	1916.82
湖　南　Hunan	472.47	156.71	25996.91	1426.56	1860.62
广　东　Guangdong	1305.85		39245.71	1214.30	1411.34
广　西　Guangxi	403.58	36.63	14701.04	345.64	534.23
海　南　Hainan	94.06		3318.31	123.59	68.07
重　庆　Chongqing	319.89	2.70	6029.37	256.50	221.47
四　川　Sichuan	791.17		27186.31	1894.27	1392.46
贵　州　Guizhou	79.67	182.50	18136.54	1082.63	388.38
云　南　Yunnan	51.51	16.37	10991.29	428.05	481.47
西　藏　Tibet	3.03	5.66	665.11	21.15	1.20
陕　西　Shaanxi	161.28	953.86	15283.86	835.28	690.83
甘　肃　Gansu	35.62	1532.16	7263.31	289.17	417.00
青　海　Qinghai	10.36	254.36	1328.13	19.59	7.95
宁　夏　Ningxia	33.92	947.90	1789.50	65.35	57.94
新　疆　Xinjiang	33.63	1263.84	7517.25	214.37	203.03
新疆生产 建设兵团　Xinjiang Production and Construction Corps	26.83	1488.60	1029.48	101.17	13.37

Gas, Central Heating, Road and Bridges of Towns(2022)

安装路灯的道路长度 Roads with Lamps	道路面积（万平方米） Surface Area of Roads (10,000 m²)	本年新增 Added This year	本年更新改造 Renewal and Upgrading during the Reported Year	桥梁座数（座） Number of Bridges (unit)	本年新增 Added This year	地区名称 Name of Regions
147222.58	318779.80	13106.74	13889.69	71992	2278	全　国
1348.51	1595.31	67.21	36.65	380	2	北　京
1770.81	2376.65	13.62	34.27	521	2	天　津
5872.04	8539.72	136.75	398.21	2096	44	河　北
1907.68	4795.45	158.22	127.58	825	47	山　西
1768.72	4681.23	88.72	60.28	614	26	内　蒙　古
2216.85	5020.95	165.50	270.82	1660	44	辽　宁
1866.59	3539.38	98.73	110.77	907	15	吉　林
1881.51	4471.38	10.35	84.46	554	3	黑　龙　江
2700.39	7098.00	102.20	126.86	4168	59	上　海
18763.64	28420.59	457.37	1038.47	7454	139	江　苏
9265.10	18018.47	511.72	927.98	8687	140	浙　江
9766.18	21612.99	1076.81	756.77	3351	168	安　徽
6743.65	11780.40	483.56	613.44	2367	39	福　建
3868.25	10964.45	767.22	833.19	2496	108	江　西
14354.77	26776.26	1014.08	1796.04	5015	173	山　东
8706.29	23372.09	1360.20	1073.71	3301	177	河　南
5812.13	17331.90	1035.39	1075.97	2978	198	湖　北
5987.94	15452.42	856.64	866.25	2813	111	湖　南
14932.90	27281.75	935.77	1082.41	5422	72	广　东
4188.33	10391.87	261.81	401.92	1623	58	广　西
1108.39	2071.20	71.17	45.69	304	14	海　南
2335.26	4023.61	166.42	105.67	1435	23	重　庆
6346.54	20026.89	1265.13	630.72	4525	180	四　川
3015.26	11312.05	682.16	225.18	1646	88	贵　州
2286.32	6316.25	282.92	200.76	1170	26	云　南
107.43	403.43	17.49	1.05	767	17	西　藏
3375.78	9133.63	444.83	465.03	2389	129	陕　西
2154.17	4537.60	273.37	286.36	1255	66	甘　肃
125.92	753.19	14.80	5.17	247	66	青　海
511.33	1129.11	76.83	50.56	171	2	宁　夏
1743.21	4605.26	127.87	150.09	788	41	新　疆
390.69	946.32	81.88	7.36	63	1	新疆生产建设兵团

3-2-5 建制镇排水和污水处理(2022年)
Drainage and Wastewater Treatment of Towns(2022)

地区名称 Name of Regions	对生活污水进行处理的建制镇 Town with Domestic Wastewater Treated		污水处理厂 Wastewater Treatment Plant		污水处理装置处理能力（万立方米/日）Treatment Capacity of Wastewater Treatment Facilities (10,000 m³/day)	排水管道长度（公里）Length of Drainage Pipelines (km)		排水暗渠长度（公里）Length of Drains (km)	
	个数（个）Number (unit)	比例（%）Rate (%)	个数（个）Number of Wastewater Treatment Plants (unit)	处理能力（万立方米/日）Treatment Capacity (10,000 m³/day)			本年新增 Added This Year		本年新增 Added This Year
全 国 National Total	14980	77.84	15177	3138.57	2576.43	218070.17	12494.43	120739.51	5632.99
北 京 Beijing	99	87.61	118	46.67	45.95	1397.47	24.22	382.35	11.60
天 津 Tianjin	112	99.12	180	20.80	10.71	2111.04	20.52	419.03	
河 北 Hebei	737	65.28	350	98.64	62.03	5312.31	268.39	2821.58	137.05
山 西 Shanxi	297	55.93	171	40.75	25.65	3216.02	298.85	2245.56	96.04
内 蒙 古 Inner Mongolia	142	32.57	103	16.07	13.12	2545.08	165.21	1094.22	25.62
辽 宁 Liaoning	209	34.15	144	38.62	29.38	2574.46	41.93	1585.45	31.50
吉 林 Jilin	190	48.72	157	46.75	41.18	2277.32	136.05	980.40	57.40
黑 龙 江 Heilongjiang	152	31.28	107	14.58	11.97	2356.82	48.29	1317.21	15.80
上 海 Shanghai	97	96.04	18	40.70	39.70	5854.17	182.44	1021.91	17.73
江 苏 Jiangsu	656	100.00	665	393.42	223.95	21710.76	653.28	9394.26	245.53
浙 江 Zhejiang	569	98.78	267	181.71	210.84	15856.24	656.07	5348.57	203.78
安 徽 Anhui	884	96.72	865	130.51	103.32	13872.23	796.69	8573.54	448.31
福 建 Fujian	558	99.29	586	127.80	234.06	8330.12	500.99	3928.65	138.49
江 西 Jiangxi	579	79.10	915	38.01	32.14	7138.91	562.44	6327.24	457.71
山 东 Shandong	1036	98.11	1050	314.79	271.76	18278.78	872.22	12056.14	466.80
河 南 Henan	736	67.65	555	128.16	118.38	13229.17	898.32	8933.84	539.36
湖 北 Hubei	688	98.01	608	170.64	108.05	12734.89	706.99	6730.89	280.31
湖 南 Hunan	980	91.59	1313	125.59	111.11	12275.50	1178.25	6501.88	447.24
广 东 Guangdong	943	94.02	969	617.31	463.68	20552.82	1462.59	11766.92	641.79
广 西 Guangxi	636	90.60	622	64.51	71.30	5084.03	196.15	4241.94	133.65
海 南 Hainan	110	70.51	80	8.90	4.55	1447.57	141.27	824.95	49.23
重 庆 Chongqing	588	99.83	771	82.17	44.30	4638.12	205.13	2261.34	73.95
四 川 Sichuan	1646	89.07	2577	189.54	146.20	12614.68	890.04	7613.20	307.55
贵 州 Guizhou	728	94.30	660	65.76	61.68	6707.15	393.77	4970.04	176.38
云 南 Yunnan	389	66.16	379	36.09	5.94	3611.87	343.84	3134.02	186.67
西 藏 Tibet	23	31.08	11	0.26	0.13	94.61	5.85	70.86	23.50
陕 西 Shaanxi	519	56.23	470	46.70	49.44	5664.23	357.70	3705.10	276.21
甘 肃 Gansu	355	44.82	246	24.50	13.83	2720.48	182.43	1329.38	87.94
青 海 Qinghai	34	32.69	20	4.30	4.75	412.07	29.51	126.57	16.29
宁 夏 Ningxia	70	90.91	68	6.04	3.10	1026.40	31.59	417.53	3.63
新 疆 Xinjiang	181	57.83	98	10.95	9.58	1701.46	165.74	606.92	35.93
新疆生产建设兵团 Xinjiang Production and Construction Corps	37	100.00	34	7.37	4.66	723.39	77.67	8.02	

3-2-6 建制镇园林绿化及环境卫生(2022年)
Landscaping and Environmental Sanitation of Towns(2022)

地区名称 Name of Regions	园林绿化(公顷) Landscaping(hectare)				环境卫生 Environmental Sanitation		
	绿化覆盖面积 Green Coverage Area	绿地面积 Area of Parks and Green Space	本年新增 Added This Year	公园绿地面积 Public Green Space	生活垃圾中转站 (座) Number of Garbage Transfer Station (unit)	环卫专用车辆设备 (辆) Number of Special Vehicles for Environmental Sanitation (unit)	公共厕所 (座) Number of Latrines (unit)
全 国 National Total	750745.44	483452.78	15532.48	49908.57	26579	115099	126443
北 京 Beijing	7339.42	5099.97	561.31	324.94	292	1150	1364
天 津 Tianjin	6895.86	4209.79	31.65	359.34	166	1063	961
河 北 Hebei	26473.69	15694.18	538.32	1060.75	950	4394	4575
山 西 Shanxi	13740.48	7271.97	170.66	504.37	510	3855	3186
内 蒙 古 Inner Mongolia	15864.40	10140.87	371.41	854.10	583	2192	3340
辽 宁 Liaoning	14261.10	7376.44	166.08	337.53	383	3282	1996
吉 林 Jilin	9184.39	5555.66	123.44	481.17	386	1689	1258
黑 龙 江 Heilongjiang	8121.49	5666.32	194.26	383.16	314	2140	1273
上 海 Shanghai	25563.47	16078.43	181.82	2554.55	297	2732	1845
江 苏 Jiangsu	82245.01	67012.14	1172.16	9768.87	1303	8129	8073
浙 江 Zhejiang	41451.05	28364.05	933.12	2601.75	1146	7373	10407
安 徽 Anhui	48758.92	27760.38	1023.46	1912.60	1219	6807	7099
福 建 Fujian	32965.61	24008.47	300.63	3199.21	740	3521	5056
江 西 Jiangxi	21397.87	14545.54	868.73	1084.31	1179	2769	4442
山 东 Shandong	96279.50	62643.28	1605.87	7320.59	1296	6364	6810
河 南 Henan	51783.11	20765.91	1309.10	2218.96	2274	7149	6387
湖 北 Hubei	36092.44	19403.27	1151.47	1561.74	1232	4739	5898
湖 南 Hunan	51499.38	34243.70	1033.72	2548.57	1474	4812	5961
广 东 Guangdong	48929.21	34001.11	701.40	4958.56	1477	9898	10208
广 西 Guangxi	16726.53	10886.83	510.58	1426.01	819	4476	2255
海 南 Hainan	4794.70	3303.44	178.19	153.01	150	1251	613
重 庆 Chongqing	9749.87	5880.45	202.19	280.39	640	2094	2947
四 川 Sichuan	18447.86	13221.11	364.40	1155.56	3300	6988	9954
贵 州 Guizhou	19246.63	11447.16	420.07	716.26	1274	4063	4798
云 南 Yunnan	6912.69	4562.84	217.57	298.52	491	2475	4900
西 藏 Tibet	306.63	68.14	4.07	4.34	53	324	439
陕 西 Shaanxi	10977.84	7415.90	344.53	747.23	1227	3950	4675
甘 肃 Gansu	9072.81	5540.86	443.62	297.40	928	2803	3072
青 海 Qinghai	1017.89	620.60	10.57	7.03	89	434	555
宁 夏 Ningxia	3293.54	2040.28	138.31	143.47	130	528	412
新 疆 Xinjiang	8172.53	6278.08	195.57	520.88	248	1287	1443
新疆生产建设兵团 Xinjiang Production and Construction Corps	3179.52	2345.61	64.20	123.40	9	368	241

3-2-7 建制镇房屋(2022年)

地区名称 Name of Regions	住宅 Residential Building						公共建筑	
	年末实有建筑面积(万平方米) Total Floor Space of Buildings (year-end) (10,000 m²)	混合结构以上 Mixed Strucure and Above	本年竣工建筑面积(万平方米) Floor Space Completed This Year (10,000 m²)	混合结构以上 Mixed Strucure and above	房地产开发 Real Estate Development	人均住宅建筑面积(平方米) Per Capita Floor Space (m²)	年末实有建筑面积(万平方米) Total Floor Space of Buildings (year-end) (10,000 m²)	混合结构以上 Mixed Strucure and Above
全 国 National Total	652391.35	534849.20	22229.47	15948.46	9096.30	39.23	174648.14	151194.70
北 京 Beijing	4434.28	3790.93	147.63	123.61	53.49	64.85	1315.05	1245.87
天 津 Tianjin	6260.65	5865.09	183.08	157.18	151.15	52.46	1287.46	1261.80
河 北 Hebei	20556.91	16733.35	299.93	212.71	117.84	31.84	3300.66	3120.11
山 西 Shanxi	10928.81	6512.14	289.81	137.40	101.43	37.30	1623.37	1248.85
内 蒙 古 Inner Mongolia	7404.46	6210.67	36.49	28.79	14.06	33.81	2799.14	2655.86
辽 宁 Liaoning	8761.47	6467.88	168.86	140.56	142.92	30.57	2693.57	2491.63
吉 林 Jilin	7959.54	6362.80	49.57	22.37	34.05	35.93	1912.30	1835.30
黑 龙 江 Heilongjiang	8641.27	7057.13	542.81	32.77	26.52	31.42	1839.51	1739.29
上 海 Shanghai	26630.76	24889.03	1125.51	817.69	890.05	73.30	11780.17	11609.93
江 苏 Jiangsu	59484.75	54291.41	2188.89	1774.36	1132.08	48.80	19409.62	18859.28
浙 江 Zhejiang	40190.91	34660.76	2099.18	1675.82	1206.43	57.28	9294.48	8791.48
安 徽 Anhui	35391.93	28566.65	1629.49	1383.52	507.50	34.24	7289.38	6743.50
福 建 Fujian	25434.10	20488.37	669.80	516.52	327.75	40.99	8970.26	8507.17
江 西 Jiangxi	21649.02	17871.19	544.80	484.77	131.11	37.23	3848.42	3597.96
山 东 Shandong	52132.78	40273.98	1953.61	1502.34	933.46	35.96	29342.92	16056.26
河 南 Henan	44004.23	36556.16	1048.12	827.75	274.17	34.16	8351.17	7709.17
湖 北 Hubei	31280.71	24046.11	760.99	620.59	212.96	36.95	6311.69	5382.13
湖 南 Hunan	39368.60	31595.15	2616.47	879.38	255.35	36.22	8285.40	7958.69
广 东 Guangdong	54769.71	46960.81	2131.32	1557.95	1403.80	43.17	10223.32	9097.20
广 西 Guangxi	18194.49	15969.53	541.81	498.83	184.22	33.19	4946.11	4721.38
海 南 Hainan	4282.19	4011.86	154.31	147.57	88.05	40.01	683.05	655.35
重 庆 Chongqing	15819.61	13949.99	193.23	174.56	95.46	38.25	2550.16	2427.88
四 川 Sichuan	37784.41	31182.53	926.53	678.81	298.60	36.98	7267.28	6679.25
贵 州 Guizhou	23426.36	13967.03	547.45	449.56	107.36	41.64	4061.22	2937.31
云 南 Yunnan	13006.47	9536.61	361.09	323.76	64.55	35.29	2642.04	2409.26
西 藏 Tibet	932.00	678.78	3.95	2.95	0.34	111.04	67.41	58.20
陕 西 Shaanxi	16181.85	13002.07	292.71	265.06	63.54	30.71	3437.53	3227.27
甘 肃 Gansu	7813.92	6095.48	206.02	189.61	62.51	32.43	6269.53	5623.73
青 海 Qinghai	1207.82	898.98	9.80	9.70		33.84	455.84	399.77
宁 夏 Ningxia	2409.18	2015.51	53.43	53.08	37.60	45.50	695.86	666.49
新 疆 Xinjiang	4061.80	2782.52	273.75	244.13	13.08	32.17	1423.26	1252.53
新疆生产建设兵团 Xinjiang Production and Construction Corps	1986.37	1558.72	179.07	14.77	164.90	70.84	270.96	224.81

Building Construction of Towns(2022)

本年竣工建筑面积（万平方米）Floor Space Completed This Year (10,000 m²)	混合结构以上 Mixed Strucure and Above	房地产开发 Real Estate Development	年末实有建筑面积（万平方米）Total Floor Space of Buildings (year-end) (10,000m²)	混合结构以上 Mixed Strucure and Above	本年竣工建筑面积（万平方米）Floor Space Completed This Year (10,000 m²)	混合结构以上 Mixed Strucure and Above	房地产开发 Real Estate Development	地区名称 Name of Regions
Public Building			生产性建筑 Industrial Building					
4621.30	4121.65	566.55	220732.42	192278.17	9776.01	8815.86	618.83	全　国
13.82	13.52	0.16	1282.66	987.93				北　京
3.46	3.41	1.70	4124.68	3942.79	39.57	38.12	0.15	天　津
67.69	60.01	22.12	5318.68	3683.84	92.58	88.11	10.38	河　北
29.07	26.97	0.01	2967.48	1510.69	25.74	23.55		山　西
26.19	24.48	0.42	2032.52	1777.17	19.35	10.80		内　蒙　古
4.17	3.93	0.45	2596.13	2070.23	46.49	44.57	0.80	辽　宁
32.95	32.67		1893.91	1658.84	15.78	15.59		吉　林
1.04	1.04	0.10	2446.49	2193.69	2.49	1.64	0.69	黑　龙　江
274.31	267.94	55.41	14045.72	13725.37	282.77	268.27	30.50	上　海
342.51	313.96	34.69	39369.16	36859.46	1821.69	1710.76	35.46	江　苏
433.37	418.72	15.53	28658.92	27145.53	1655.15	1633.32	64.33	浙　江
295.28	267.83	28.77	9480.32	7688.50	561.67	516.77	11.84	安　徽
175.62	158.14	26.91	7733.98	6478.01	246.22	167.89	29.15	福　建
235.89	213.63	33.06	5232.01	4825.73	362.86	312.93	42.91	江　西
477.47	375.28	164.12	28372.89	24740.45	1057.23	886.71	115.08	山　东
193.29	175.10	14.60	8884.28	7467.26	326.84	212.83	67.42	河　南
376.39	353.93	44.44	6117.19	4679.99	371.57	325.63	51.05	湖　北
342.19	238.15	12.33	5483.43	4522.03	182.04	166.92	3.61	湖　南
397.43	352.41	67.46	19347.04	15591.94	1744.48	1635.62	116.85	广　东
82.20	77.81	3.76	2933.70	2320.39	126.85	111.71	0.86	广　西
10.19	9.57	0.27	340.98	326.95	8.13	3.61	4.50	海　南
35.73	34.04	1.96	2974.51	1975.04	48.81	44.61	0.01	重　庆
239.15	216.69	17.77	5546.16	4607.06	257.44	163.22	4.12	四　川
76.98	54.04	4.79	4746.44	4341.77	175.73	156.74	2.06	贵　州
67.76	60.37	0.92	2277.85	1811.74	59.12	48.95	0.34	云　南
8.48	8.45		38.17	31.97	0.03	0.03		西　藏
110.90	102.84	1.03	2042.11	1749.49	77.27	70.27	0.96	陕　西
185.38	176.37	3.79	2348.35	1822.78	44.38	42.50		甘　肃
3.96	3.91		175.81	150.99	0.34	0.26		青　海
13.14	13.03	3.58	711.73	637.32	17.71	17.12	2.04	宁　夏
58.28	57.22	5.45	1072.03	907.87	98.19	96.16	21.68	新　疆
7.03	6.20	0.95	137.10	45.35	7.51	0.67	2.04	新疆生产建设兵团

3-2-8 建制镇建设投入(2022年)

计量单位: 万元

地区名称 Name of Regions	本年建设投入合计 Total Construction Input of This Year	房屋 Building					小　计 Input of Municipal Public Facilities	供　水 Water Supply
		小　计 Input of House	房地产开发 Real Estate Development	住　宅 Residential Building	公共建筑 Public Building	生产性建筑 Industrial Building		
全　国　National Total	91401630	74606428	29314541	44922991	9019207	20664230	16795202	1457050
北　京　Beijing	689631	631946	367584	565569	65420	957	57685	1468
天　津　Tianjin	1608441	1368182	1115862	1183829	39123	145230	240260	2300
河　北　Hebei	811852	639530	152628	422056	87255	130219	172322	10158
山　西　Shanxi	639045	445247	159683	370354	43787	31106	193798	8212
内　蒙古　Inner Mongolia	237624	144456	34794	68380	47964	28112	93168	8402
辽　宁　Liaoning	530573	398216	259239	229910	13627	154680	132357	9108
吉　林　Jilin	338392	161259	61551	88307	49680	23271	177133	11470
黑龙江　Heilongjiang	174147	107893	4899	102391	2716	2787	66254	3004
上　海　Shanghai	10530449	9724644	4667548	7177770	1313626	1233248	805805	11250
江　苏　Jiangsu	12013942	10281025	2937186	4995670	706453	4578902	1732917	129629
浙　江　Zhejiang	13058320	10935120	4868619	6060978	970951	3903190	2123200	126834
安　徽　Anhui	4434215	3338812	960177	2192215	400545	746051	1095403	96096
福　建　Fujian	2307429	1755841	825771	1116401	263894	375545	551589	57447
江　西　Jiangxi	1673226	1030315	238183	582156	234999	213160	642912	86646
山　东　Shandong	7380961	5735254	1572789	3337138	740640	1657476	1645707	162341
河　南　Henan	2374179	1767263	547932	1247969	235348	283947	606916	55777
湖　北　Hubei	2674253	1938139	322198	1283834	364226	290079	736114	80958
湖　南　Hunan	2217564	1478703	296285	1019041	274582	185080	738862	94151
广　东　Guangdong	16058343	14389422	7412492	7208871	1937618	5242932	1668921	143010
广　西　Guangxi	1265638	1016166	208307	700267	104683	211215	249472	33067
海　南　Hainan	401411	208920	64082	187279	14143	7498	192492	17191
重　庆　Chongqing	628493	390241	158204	264363	59347	66531	238252	25315
四　川　Sichuan	3098653	2104510	506036	1390614	291185	422711	994143	126508
贵　州　Guizhou	1461921	1159297	293290	795812	134740	228745	302624	36823
云　南　Yunnan	1139387	800528	218283	572687	126846	100995	338859	44428
西　藏　Tibet	167680	158426	1005	141947	14474	2004	9254	1210
陕　西　Shaanxi	1212002	651358	135105	374978	175001	101379	560644	37178
甘　肃　Gansu	1079102	885639	549627	628673	139164	117802	193463	23551
青　海　Qinghai	54017	24618	375	20536	3636	446	29399	319
宁　夏　Ningxia	232787	160810	116216	112752	24795	23262	71978	3629
新　疆　Xinjiang	557555	477294	172013	293227	82274	101794	80261	7758
新疆生产建设兵团　Xinjiang Production and Construction Corps	350396	297359	86581	187017	56465	53877	53037	1814

Construction Input of Towns(2022)

Measurement Unit: 10,000 RMB

市政公用设施 Municipal Public Facilities					园林绿化	环境卫生		其 他	地区名称
燃 气 Gas Supply	集中供热 Central Heating	道路桥梁 Road and Bridge	排 水 Drainage	污水处理 Wastewater Treatment	Landscaping	环境卫生 Environmental Sanitation	垃圾处理 Garbage Treatment	Other	Name of Regions
727115	**495510**	**5469059**	**3796774**	**2599341**	**1496821**	**2046142**	**1118598**	**1306731**	全　　国
5847	8970	17023	7979	7323	5395	5955	2786	5048	北　　京
694	7929	31104	14543	10968	145946	6971	2830	30773	天　　津
23874	23144	34550	30294	21698	13029	33283	19756	3990	河　　北
9009	60313	36964	46553	37382	9370	19896	7823	3482	山　　西
1155	25742	13036	21141	17607	7563	12862	8819	3268	内 蒙 古
1794	42314	47748	7910	4463	2878	15026	9641	5578	辽　　宁
2497	25260	53387	45919	35562	7085	12086	6421	19430	吉　　林
43	13307	9267	18723	14864	2385	9957	5601	9567	黑 龙 江
13059	87	356040	196916	69850	56741	92835	49401	78878	上　　海
113286	3364	587424	335325	224038	205448	240645	116555	117795	江　　苏
59609	565	921506	382989	257172	209096	248924	101716	173677	浙　　江
44773	570	339688	273828	163312	118799	145287	79871	76363	安　　徽
11219	749	189999	128171	100282	44872	89455	62599	29677	福　　建
20023	1297	183755	172755	110506	45486	63446	32673	69504	江　　西
123223	149129	438019	290669	203575	159815	211944	101184	110567	山　　东
54885	9834	172975	112687	63502	82133	87617	41682	31007	河　　南
23689	1616	238174	198006	133433	69516	71671	43520	52485	湖　　北
26591	955	148979	283879	218235	37544	92909	52123	53854	湖　　南
25465		687377	418821	311151	97183	183201	105674	113863	广　　东
4695	11	83059	69487	44226	15047	33506	26371	10600	广　　西
2939		35734	59883	53745	1957	71877	67869	2910	海　　南
8317	30	68849	64156	54462	19975	32254	18634	19358	重　　庆
85208		317229	236485	179074	48450	92797	58412	87466	四　　川
8480	525	94947	84077	68106	9581	37536	24633	30654	贵　　州
2477	10	91023	108609	87529	13695	33192	21891	45426	云　　南
0	245	2499	2091	1911	882	2248	1448	80	西　　藏
41889	34392	138170	111965	63562	47498	58787	27971	90765	陕　　西
1903	49872	44266	29180	17867	9799	22735	12525	12159	甘　　肃
90	13	9243	11775	986	541	1241	700	6178	青　　海
3865	1825	43594	7830	5017	2496	6143	2790	2597	宁　　夏
6059	22775	12484	18053	13620	3219	7372	3824	2542	新　　疆
454	10666	20948	6076	4312	3401	2486	855	7192	新疆生产 建设兵团

3-2-9 乡市政公用设施水平(2022年)

地区名称 Name of Regions	人口密度 (人／平方公里) Population Density (person/km^2)	人均日生 活用水量 (升) Per Capita Daily Water Consumption (liter)	供 水 普及率 (%) Water Coverage Rate (%)	燃 气 普及率 (%) Gas Coverage Rate (%)	人均道路 面 积 (平方米) Road Surface Area Per Capita (m^2)	排水管道 暗渠密度 (公里／平方公里) Density of Drains (km/km^2)
全 国 **National Total**	**3612**	**99.46**	**84.72**	**33.54**	**23.95**	**7.33**
北 京 Beijing	2817	142.38	98.07	62.27	24.36	5.14
天 津 Tianjin	793	81.77	84.38	71.96	19.56	2.38
河 北 Hebei	3013	91.35	83.64	61.94	17.32	4.05
山 西 Shanxi	3188	85.12	75.63	19.84	19.54	4.85
内 蒙 古 Inner Mongolia	1819	86.94	74.40	18.16	28.55	3.24
辽 宁 Liaoning	3395	105.91	57.38	14.41	26.79	4.71
吉 林 Jilin	2545	89.20	87.48	32.76	25.20	4.47
黑 龙 江 Heilongjiang	2191	83.93	89.08	8.23	24.77	3.58
上 海 Shanghai	2320	116.38	100.00	78.00	41.53	14.30
江 苏 Jiangsu	5591	91.02	99.53	96.01	25.02	15.74
浙 江 Zhejiang	2961	109.98	85.88	50.48	30.63	10.73
安 徽 Anhui	3649	103.08	91.10	45.83	25.11	10.40
福 建 Fujian	4488	104.52	93.75	64.28	23.82	11.20
江 西 Jiangxi	4223	100.00	84.51	43.34	22.55	10.30
山 东 Shandong	3379	91.83	95.29	63.44	23.98	9.24
河 南 Henan	4254	107.34	85.27	32.90	22.21	6.78
湖 北 Hubei	3373	101.95	84.96	47.39	26.17	8.56
湖 南 Hunan	3502	110.74	75.40	32.95	21.60	6.34
广 东 Guangdong	3197	77.31	100.00	52.67	25.91	8.62
广 西 Guangxi	5503	106.21	94.12	65.22	25.34	9.53
海 南 Hainan	2328	92.88	96.39	90.19	33.34	6.87
重 庆 Chongqing	4680	88.34	90.09	43.57	18.40	12.43
四 川 Sichuan	4278	101.18	86.76	31.32	18.99	8.34
贵 州 Guizhou	3702	104.27	84.57	11.74	31.97	11.51
云 南 Yunnan	4543	94.77	94.08	9.93	19.96	10.60
西 藏 Tibet	4493	95.84	40.06	12.57	44.14	7.97
陕 西 Shaanxi	3083	103.67	78.29	17.02	28.76	7.68
甘 肃 Gansu	3387	74.95	93.05	10.60	20.25	7.63
青 海 Qinghai	4157	85.82	65.14	3.36	19.08	5.10
宁 夏 Ningxia	3054	99.52	95.52	15.88	30.11	7.64
新 疆 Xinjiang	2981	96.39	92.12	12.13	36.54	4.87

Level of Municipal Public Facilities of Built-up Area of Townships(2022)

污水处理率 (%) Wastewater Treatment Rate (%)	污水处理厂 集中处理率 Centralized Treatment Rate of Wastewater Treatment Plants	人均公园 绿地面积 (平方米) Public Recreational Green Space Per Capita (m²)	绿化 覆盖率 (%) Green Coverage Rate (%)	绿地率 (%) Green Space Rate (%)	生活垃圾 处理率 (%) Domestic Garbage Treatment Rate (%)	无害化 处理率 Domestic Garbage Harmless Treatment Rate	地区名称 Name of Regions
28.29	18.96	1.82	15.16	8.72	82.99	62.30	全　国
36.10	31.55	4.05	24.46	16.97	100.00	99.35	北　京
87.29	87.29		6.74	2.68	100.00	100.00	天　津
41.80	22.78	1.23	13.53	9.01	99.15	96.89	河　北
11.95	7.48	1.22	23.85	8.46	51.88	18.23	山　西
7.10	6.81	1.49	12.10	7.79	37.59	22.63	内　蒙　古
4.88	4.47	0.47	14.77	7.09	46.84	27.66	辽　宁
5.81	5.47	1.90	12.03	9.05	97.40	90.51	吉　林
2.87	1.60	0.82	7.95	5.30	55.22	47.15	黑　龙　江
74.77	74.77	1.14	26.27	22.03	100.00	100.00	上　海
74.89	65.92	6.03	31.02	24.32	99.90	96.18	江　苏
52.15	19.20	2.82	12.53	7.11	93.15	85.10	浙　江
55.00	46.68	2.96	18.58	11.10	99.26	97.96	安　徽
83.52	62.36	6.66	24.20	15.75	99.99	99.99	福　建
31.82	22.65	1.22	13.39	8.59	90.64	65.11	江　西
53.05	33.52	2.41	23.28	13.27	99.98	97.27	山　东
20.05	14.26	1.83	17.13	7.52	85.84	62.81	河　南
76.07	46.79	1.73	12.32	6.37	98.49	90.72	湖　北
13.86	7.49	2.47	19.38	11.68	81.54	53.87	湖　南
74.56	49.41	1.91	23.07	14.80	54.06	48.97	广　东
12.18	9.88	4.08	16.16	11.45	94.98	56.91	广　西
73.11	53.31	0.29	20.54	14.22	99.74	99.68	海　南
82.70	67.05	0.57	12.93	7.92	92.18	61.93	重　庆
37.87	31.83	2.11	9.47	6.57	85.16	64.07	四　川
35.21	28.38	0.89	11.94	6.75	83.59	50.61	贵　州
27.15	12.41	0.93	8.87	5.61	73.82	43.35	云　南
0.65	0.30	0.09	10.09	5.62	64.20	17.40	西　藏
31.02	20.79	0.06	5.97	4.51	90.95	17.22	陕　西
22.86	19.62	0.57	13.96	9.00	64.67	54.17	甘　肃
2.71	2.23	0.00	8.71	5.50	58.09	20.71	青　海
47.39	39.15	1.12	14.43	8.94	96.76	70.80	宁　夏
11.13	4.90	1.49	18.34	13.07	71.33	43.23	新　疆

3-2-10 乡基本情况(2022年)

地区名称 Name of Regions	乡个数 (个) Number of Townships (unit)	建成区 面 积 (公顷) Surface Area of Built-up Districts (hectare)	建成区 户籍人口 (万人) Registered Permanent Population (10,000 persons)	建成区 常住人口 (万人) Permanant Population (10,000 persons)
全 国 National Total	7959	568547.20	2124.25	2053.35
北 京 Beijing	15	667.94	2.34	1.88
天 津 Tianjin	3	1084.25	0.74	0.86
河 北 Hebei	580	45320.58	150.83	136.57
山 西 Shanxi	430	29839.13	92.46	95.12
内 蒙 古 Inner Mongolia	263	22054.71	43.52	40.11
辽 宁 Liaoning	187	11012.74	38.67	37.39
吉 林 Jilin	163	12602.58	33.59	32.08
黑 龙 江 Heilongjiang	318	25544.05	69.32	55.97
上 海 Shanghai	2	136.38	0.24	0.32
江 苏 Jiangsu	15	2346.30	13.79	13.12
浙 江 Zhejiang	245	15815.30	56.33	46.84
安 徽 Anhui	218	23142.23	87.48	84.45
福 建 Fujian	243	15712.93	78.94	70.52
江 西 Jiangxi	535	42373.68	190.97	178.94
山 东 Shandong	56	7280.07	25.71	24.60
河 南 Henan	570	90440.04	383.20	384.75
湖 北 Hubei	152	24085.40	73.89	81.25
湖 南 Hunan	381	39708.79	137.35	139.08
广 东 Guangdong	11	728.00	2.95	2.33
广 西 Guangxi	307	12530.52	79.31	68.96
海 南 Hainan	21	891.45	2.09	2.08
重 庆 Chongqing	162	5621.46	28.80	26.31
四 川 Sichuan	620	15733.16	67.14	67.30
贵 州 Guizhou	301	24733.55	97.51	91.57
云 南 Yunnan	537	32301.86	138.52	146.75
西 藏 Tibet	521	7825.42	33.27	35.16
陕 西 Shaanxi	18	1088.97	3.78	3.36
甘 肃 Gansu	330	12489.60	44.49	42.30
青 海 Qinghai	220	6422.08	27.90	26.70
宁 夏 Ningxia	87	5332.78	18.63	16.29
新 疆 Xinjiang	448	33681.25	100.49	100.42

Summary of Townships(2022)

设有村镇建设管理机构的个数（个）Number of Towns with Construction Management Institution (unit)	村镇建设管理人员（人）Number of Construction Management Personnel (person)	规划建设管理 Planning and Administer		有总体规划的乡个数（个）Number of Townships with Master Plans (unit)	本年编制 Compiled This Year	本年规划编制投入（万元）Input in Planning This Year (10,000 RMB)	地区名称 Name of Regions
		专职人员 Full-time Staff					
6312	18652	11681		6026	316	71068.93	全　国
15	76	39		12		300.00	北　京
3	19	10				38.30	天　津
460	968	648		347	24	1145.75	河　北
312	842	598		201	3	435.00	山　西
212	466	298		203	6	1154.95	内　蒙　古
184	251	216		155	3	901.00	辽　宁
157	316	231		91	3	693.00	吉　林
304	530	328		232	17	64.50	黑　龙　江
2	22	10		2			上　海
15	92	67		15	3	442.00	江　苏
199	548	349		216	12	1460.30	浙　江
187	592	359		183	15	4099.82	安　徽
216	468	318		227	11	6124.19	福　建
509	1624	1024		507	29	7392.62	江　西
55	183	133		50	6	156.00	山　东
553	2834	1820		451	14	3511.03	河　南
151	649	412		135	3	9152.00	湖　北
325	1102	589		303	13	7687.18	湖　南
9	44	15		11		284.00	广　东
304	1029	659		283	4	1336.91	广　西
21	44	31		21		193.44	海　南
162	416	309		151	10	425.80	重　庆
287	623	377		257	10	1093.91	四　川
294	606	370		280	31	1082.70	贵　州
510	1783	1076		494	28	12528.96	云　南
101	533	332		275	40	538.05	西　藏
12	23	14		12	1	12.00	陕　西
221	630	328		268	6	2621.42	甘　肃
128	145	90		157	16	283.00	青　海
81	160	91		77	4	708.50	宁　夏
323	1034	540		410	4	5202.60	新　疆

3-2-11 乡供水(2022年)

地区名称 Name of Regions	集中供水 的乡个数 （个） Number of Townships with Access to Piped Water (unit)	占全部乡 的比例 （%） Percentage of Total Rate （%）	公共供水综合 生产能力 （万立方米／日） Integrated Production Capacity of Public Water Supply Facilities (10,000 m³/day)	自备水源单位 综合生产能力 （万立方米／日） Integrated Production Capacity of Self-built Water Supply Facilities (10,000 m³/day)	年供水总量 （万立方米） Annual Supply of Water (10,000 m³)
全　国　National Total	7342	92.25	1242.84	420.67	127837.76
北　京　Beijing	15	100.00	2.68	0.80	161.21
天　津　Tianjin	3	100.00	0.36	0.20	43.86
河　北　Hebei	524	90.34	44.98	24.30	7157.51
山　西　Shanxi	415	96.51	29.64	9.03	4742.62
内　蒙　古　Inner Mongolia	234	88.97	16.98	2.79	1877.67
辽　宁　Liaoning	118	63.10	7.63	3.00	1511.48
吉　林　Jilin	163	100.00	20.11	10.65	1651.93
黑　龙　江　Heilongjiang	304	95.60	27.02	50.74	2203.21
上　海　Shanghai	2	100.00	0.96	0.09	47.01
江　苏　Jiangsu	15	100.00	5.27	0.72	769.63
浙　江　Zhejiang	237	96.73	27.13	10.98	3695.21
安　徽　Anhui	217	99.54	38.70	10.00	4890.01
福　建　Fujian	242	99.59	33.03	8.95	5097.61
江　西　Jiangxi	528	98.69	123.21	50.77	10341.88
山　东　Shandong	55	98.21	10.86	3.62	1665.73
河　南　Henan	566	99.30	192.77	99.16	24477.50
湖　北　Hubei	150	98.68	63.17	20.80	5071.35
湖　南　Hunan	367	96.33	122.41	38.56	11670.47
广　东　Guangdong	10	90.91	1.15	0.57	115.47
广　西　Guangxi	304	99.02	41.45	4.15	3653.73
海　南　Hainan	21	100.00	1.17	0.12	113.13
重　庆　Chongqing	162	100.00	7.22	1.54	1273.16
四　川　Sichuan	619	99.84	55.99	15.80	4080.63
贵　州　Guizhou	287	95.35	24.61	7.95	5215.91
云　南　Yunnan	536	99.81	72.13	31.45	10262.65
西　藏　Tibet	222	42.61	108.35	4.95	6706.46
陕　西　Shaanxi	17	94.44	0.76	0.16	135.34
甘　肃　Gansu	326	98.79	28.93	1.75	1852.64
青　海　Qinghai	152	69.09	4.58	0.57	866.19
宁　夏　Ningxia	87	100.00	8.24	1.41	909.57
新　疆　Xinjiang	444	99.11	121.35	5.09	5576.99

Water Supply of Townships(2022)

年生活用水量 Annual Domestic Water Consumption	年生产用水量 Annual Water Consumption for Production	供水管道长度（公里） Length of Water Supply Pipelines (km)	本年新增 Added This Year	用水人口（万人）Population with Access to Water (10,000 persons)	地区名称 Name of Regions
63150.11	53083.34	148490.00	6542.85	1739.57	全　国
95.91	32.61	195.50	4.00	1.85	北　京
21.65	20.95	47.31		0.73	天　津
3808.35	2819.36	5117.37	153.86	114.22	河　北
2235.17	2031.42	7649.80	147.32	71.94	山　西
946.87	778.35	2273.27	81.35	29.84	内　蒙　古
829.47	575.13	1780.90	82.01	21.46	辽　宁
913.52	512.42	2027.93	52.70	28.06	吉　林
1527.40	599.49	2909.50	19.85	49.86	黑　龙　江
13.44	9.98	39.00		0.32	上　海
433.82	323.31	547.42	5.20	13.06	江　苏
1614.77	1639.69	4192.22	193.30	40.23	浙　江
2894.46	1519.99	5607.65	280.04	76.93	安　徽
2522.00	1901.69	5175.77	214.80	66.11	福　建
5520.30	3783.46	9821.30	699.07	151.23	江　西
785.70	820.13	1239.59	73.42	23.44	山　东
12853.82	8961.16	21940.77	1086.51	328.07	河　南
2568.69	2058.57	6871.98	357.52	69.03	湖　北
4238.29	6712.43	12281.20	476.17	104.86	湖　南
65.68	46.34	82.44	3.24	2.33	广　东
2516.02	907.09	3385.48	166.62	64.90	广　西
67.80	41.26	154.64	5.60	2.00	海　南
764.18	410.30	1670.99	46.11	23.70	重　庆
2156.57	1331.80	9055.81	465.30	58.39	四　川
2947.06	1484.89	9088.13	531.55	77.43	贵　州
4775.94	4746.57	15873.38	923.93	138.07	云　南
492.69	6037.14	2555.13	138.92	14.08	西　藏
99.47	21.53	166.23	3.32	2.63	陕　西
1076.71	615.41	2312.78	58.45	39.36	甘　肃
544.80	308.78	1131.31	2.92	17.39	青　海
565.09	291.94	1191.92	65.96	15.56	宁　夏
3254.47	1740.15	12103.28	203.81	92.51	新　疆

3-2-12 乡燃气、供热、道路桥梁(2022年)

地区名称 Name of Regions		用气人口 (万人) Population with Access to Gas (10,000 persons)	集中供热 (万平方米) Area of Centrally Heated District (10,000 m²)	道路长度 (公里) Length of Roads (km)	本年新增 Added This Year	本年更 新改造 Renewal and Upgrading during The Reported Year
全 国	National Total	688.71	2645.43	88532.31	4248.00	4546.85
北 京	Beijing	1.17	26.63	87.86	3.00	2.00
天 津	Tianjin	0.62	4.00	38.52		2.00
河 北	Hebei	84.59	423.04	4717.73	71.87	202.26
山 西	Shanxi	18.88	271.79	3289.94	98.48	141.08
内 蒙 古	Inner Mongolia	7.28	180.94	2143.96	37.80	31.78
辽 宁	Liaoning	5.39	162.15	1929.01	133.00	79.92
吉 林	Jilin	10.51	129.01	1513.06	52.63	105.23
黑 龙 江	Heilongjiang	4.61	142.39	2656.22	8.75	52.76
上 海	Shanghai	0.25		30.31		1.20
江 苏	Jiangsu	12.60		563.24	2.90	14.90
浙 江	Zhejiang	23.64	1.00	2646.38	95.33	190.37
安 徽	Anhui	38.70		3647.27	228.61	167.38
福 建	Fujian	45.33		3064.90	107.60	121.04
江 西	Jiangxi	77.55	124.97	7014.35	471.40	598.76
山 东	Shandong	15.61	129.87	925.12	48.14	71.90
河 南	Henan	126.59	228.02	15078.78	1008.04	670.68
湖 北	Hubei	38.50	33.66	3789.28	216.51	238.49
湖 南	Hunan	45.82	72.60	5189.45	281.94	543.71
广 东	Guangdong	1.23		142.05	24.47	13.49
广 西	Guangxi	44.98	1.00	3041.56	98.05	80.65
海 南	Hainan	1.87		129.20	5.99	2.00
重 庆	Chongqing	11.46		867.13	20.62	63.53
四 川	Sichuan	21.08		2532.18	160.24	88.91
贵 州	Guizhou	10.75	13.99	5058.47	266.91	144.30
云 南	Yunnan	14.57	8.82	5539.06	325.31	341.08
西 藏	Tibet	4.42	34.18	2975.38	104.87	72.16
陕 西	Shaanxi	0.57	10.90	194.35	8.10	10.70
甘 肃	Gansu	4.48	121.24	1619.84	26.08	81.01
青 海	Qinghai	0.90	54.50	966.83	30.80	9.41
宁 夏	Ningxia	2.59	94.47	837.90	42.57	17.48
新 疆	Xinjiang	12.18	376.26	6302.98	267.99	386.67

Gas, Central Heating, Roads and Bridges of Townships(2022)

安装路灯的道路长度 Roads with Lamps	道路面积（万平方米） Surface Area of Roads (10,000 m²)	本年新增 Added This year	本年更新改造 Renewal and Upgrading during The Reported Year	桥梁座数（座） Number of Bridges (unit)	本年新增 Added This year	地区名称 Name of Regions
21156.06	49173.02	2851.20	2357.37	14998	734	全　国
54.20	45.84	2.10	1.00	21		北　京
17.00	16.82		0.80	2		天　津
1694.99	2365.61	51.14	126.89	792	26	河　北
825.42	1858.99	71.50	74.77	430	7	山　西
427.76	1145.07	23.87	29.74	210	9	内 蒙 古
381.92	1001.78	49.90	35.35	402	17	辽　宁
433.09	808.21	27.22	52.68	246	5	吉　林
746.44	1386.31	8.06	27.59	226	4	黑 龙 江
11.56	13.14		0.60	10		上　海
252.05	328.19	3.71	16.42	100	5	江　苏
562.12	1434.74	98.82	82.12	873	15	浙　江
1212.07	2120.31	161.01	110.82	846	52	安　徽
906.79	1679.50	74.57	72.91	713	14	福　建
1546.94	4035.25	323.49	300.81	1337	88	江　西
275.76	589.94	27.53	33.50	231	11	山　东
3327.35	8543.56	591.31	359.43	2012	103	河　南
951.41	2126.08	134.12	132.81	390	47	湖　北
1019.14	3004.01	207.48	261.14	1064	49	湖　南
26.35	60.31	8.83	5.65	26	1	广　东
888.60	1747.79	63.26	41.14	479	30	广　西
44.98	69.18	1.29	1.29	31	5	海　南
292.45	484.00	16.12	16.36	278	11	重　庆
450.10	1278.14	80.54	42.75	718	42	四　川
922.56	2927.35	289.17	92.25	376	25	贵　州
1198.90	2928.68	163.16	157.23	864	31	云　南
380.72	1551.84	109.98	36.29	1002	80	西　藏
14.30	96.57	16.02	12.25	11		陕　西
360.97	856.48	18.26	39.28	337	21	甘　肃
125.16	509.43	19.54	4.65	244	5	青　海
289.90	490.41	39.64	22.35	123	5	宁　夏
1515.06	3669.49	169.56	166.50	604	26	新　疆

3-2-13 乡排水和污水处理(2022年)
Drainage and Wastewater Treatment of Townships(2022)

地区名称 Name of Regions	对生活污水进行处理的乡 Township with Domestic Wastewater Treated 个数(个) Number (unit)	比例(%) Rate (%)	污水处理厂 Wastewater Treatment Plant 个数(个) Number of Wastwater Treatment Plants (unit)	处理能力(万立方米/日) Treatment Capacity (10,000 m³/day)	污水处理装置处理能力(万立方米/日) Treatment Capacity of Wastewater Treatment Facilities (10,000 m³/day)	排水管道长度(公里) Length of Drainage Pipelines (km)	本年新增 Added This Year	排水暗渠长度(公里) Length of Drains (km)	本年新增 Added This Year
全 国 National Total	3636	45.68	2456	146.08	127.83	22966.30	1733.46	18727.58	997.25
北 京 Beijing	8	53.33	3	0.18	0.82	20.80	1.60	13.50	
天 津 Tianjin	3	100.00	3	0.06	0.09	22.75		3.06	
河 北 Hebei	295	50.86	43	4.95	3.08	1159.57	55.40	675.33	32.67
山 西 Shanxi	82	19.07	21	3.56	3.82	811.95	46.60	634.45	17.74
内蒙古 Inner Mongolia	28	10.65	8	0.07	0.04	389.57	21.10	324.49	19.32
辽 宁 Liaoning	24	12.83	11	0.28	0.27	229.09	0.80	289.82	2.00
吉 林 Jilin	30	18.40	12	0.37	0.50	295.63	20.66	267.35	8.73
黑龙江 Heilongjiang	26	8.18	11	0.14	0.14	439.62	11.50	475.99	
上 海 Shanghai	2	100.00	2	0.11	0.11	13.50		6.00	
江 苏 Jiangsu	15	100.00	12	2.94	0.84	247.97	6.43	121.31	0.03
浙 江 Zhejiang	238	97.14	47	3.70	4.42	997.12	31.37	699.55	46.58
安 徽 Anhui	205	94.04	191	11.07	10.72	1355.56	104.19	1052.31	41.06
福 建 Fujian	242	99.59	298	15.60	12.90	1140.49	49.09	619.91	26.75
江 西 Jiangxi	319	59.63	267	7.99	13.07	2364.26	210.93	2001.86	165.77
山 东 Shandong	44	78.57	34	4.12	4.24	401.30	25.79	271.19	11.78
河 南 Henan	295	51.75	147	34.77	32.64	3474.57	277.00	2660.42	161.74
湖 北 Hubei	150	98.68	105	9.58	8.59	1179.83	90.83	880.78	54.11
湖 南 Hunan	195	51.18	105	5.07	3.61	1452.36	122.54	1065.60	72.59
广 东 Guangdong	10	90.91	24	0.93	0.56	27.94		34.81	
广 西 Guangxi	76	24.76	58	1.64	1.46	669.62	43.19	524.80	50.19
海 南 Hainan	16	76.19	14	0.27	0.21	36.22	3.46	25.03	3.90
重 庆 Chongqing	157	96.91	168	4.68	3.08	397.79	14.61	301.17	13.63
四 川 Sichuan	318	51.29	275	10.89	9.68	719.01	57.38	593.42	21.83
贵 州 Guizhou	185	61.46	152	9.04	7.52	1117.34	106.96	1728.84	59.74
云 南 Yunnan	287	53.45	267	9.62	1.57	1667.37	168.26	1757.46	120.79
西 藏 Tibet	38	7.29	15	0.27	0.09	214.79	30.96	409.08	11.38
陕 西 Shaanxi	9	50.00	4	0.07	0.08	39.69	0.61	43.94	1.77
甘 肃 Gansu	110	33.33	52	1.25	1.44	481.05	18.75	472.15	22.60
青 海 Qinghai	17	7.73	5	0.35	0.36	190.88	7.10	136.40	5.18
宁 夏 Ningxia	54	62.07	49	1.44	0.71	269.20	12.36	138.05	16.84
新 疆 Xinjiang	158	35.27	53	1.06	1.18	1139.46	193.99	499.51	8.53

3-2-14 乡园林绿化及环境卫生(2022年)
Landscaping and Environmental Sanitation of Townships(2022)

地区名称 Name of Regions	园林绿化(公顷) Landscaping(hectare)				环境卫生 Environmental Sanitation		
	绿化覆盖面积 Green Coverage Area	绿地面积 Area of Parks and Green Space	本年新增 Added This Year	公园绿地面积 Public Green Space	生活垃圾中转站(座) Number of Garbage Transfer Station (unit)	环卫专用车辆设备(辆) Number of Special Vehicles for Environmental Sanitation (unit)	公共厕所(座) Number of Latrines (unit)
全 国 National Total	86213.36	49561.45	2452.74	3727.91	8583	27005	35619
北 京 Beijing	163.38	113.38	2.58	7.63	8	74	117
天 津 Tianjin	73.12	29.07			7	15	17
河 北 Hebei	6129.61	4084.52	185.59	167.48	462	1697	2241
山 西 Shanxi	7115.47	2523.81	50.90	116.15	252	1907	1560
内 蒙 古 Inner Mongolia	2668.33	1717.96	20.80	59.83	187	1012	1238
辽 宁 Liaoning	1626.43	781.03	28.80	17.76	102	740	488
吉 林 Jilin	1516.57	1139.93	52.59	61.03	125	540	346
黑 龙 江 Heilongjiang	2030.58	1352.79	82.47	45.92	117	1013	546
上 海 Shanghai	35.83	30.05		0.36	4	16	4
江 苏 Jiangsu	727.87	570.69	7.70	79.15	20	104	110
浙 江 Zhejiang	1981.20	1124.50	91.25	131.86	314	910	1871
安 徽 Anhui	4299.51	2568.13	121.05	249.86	240	1175	1385
福 建 Fujian	3802.36	2475.04	82.84	469.97	246	747	1398
江 西 Jiangxi	5674.06	3638.06	201.74	217.65	1168	1612	2538
山 东 Shandong	1694.70	966.25	33.97	59.25	54	216	251
河 南 Henan	15495.00	6800.50	351.78	704.17	1116	2895	3147
湖 北 Hubei	2966.37	1533.26	114.49	140.18	199	773	970
湖 南 Hunan	7695.00	4637.57	345.15	344.01	752	1046	1813
广 东 Guangdong	167.96	107.71	1.36	4.44	8	58	81
广 西 Guangxi	2025.53	1434.71	70.54	281.20	266	1031	685
海 南 Hainan	183.08	126.75	4.88	0.60	17	93	71
重 庆 Chongqing	727.13	445.48	15.24	14.89	164	376	452
四 川 Sichuan	1489.92	1034.01	30.24	142.14	682	1248	2224
贵 州 Guizhou	2953.72	1668.50	44.59	81.21	297	1142	1509
云 南 Yunnan	2864.88	1811.62	119.45	136.00	327	1501	3706
西 藏 Tibet	789.88	440.02	46.63	3.07	348	1229	3001
陕 西 Shaanxi	64.96	49.10	3.60	0.19	15	49	114
甘 肃 Gansu	1743.47	1124.68	102.38	24.21	522	885	951
青 海 Qinghai	559.39	353.05	15.76	0.01	126	908	796
宁 夏 Ningxia	769.44	476.84	41.59	18.19	97	406	291
新 疆 Xinjiang	6178.61	4402.44	182.78	149.50	341	1587	1698

3-2-15　乡房屋(2022年)

地区名称 Name of Regions	住宅　Residential Building						公共建筑	
	年末实有建筑面积 (万平方米) Total Floor Space of Buildings (year-end) (10,000 m²)	混合结构以上 Mixed Strucure and Above	本年竣工建筑面积 (万平方米) Floor Space Completed This Year (10,000 m²)	混合结构以上 Mixed Strucure and Above	房地产开发 Real Estate Development	人均住宅建筑面积 (平方米) Per Capita Floor Space (m²)	年末实有建筑面积 (万平方米) Total Floor Space of Buildings (year-end) (10,000 m²)	混合结构以上 Mixed Strucure and Above
全　国 National Total	77480.67	58620.18	1824.97	1505.14	292.64	36.47	20533.94	16135.79
北　京 Beijing	142.15	142.01	3.03	3.03		60.63	16.71	16.00
天　津 Tianjin	37.22	21.62				50.38	4.15	3.73
河　北 Hebei	4683.91	3794.21	117.70	74.98	12.47	31.05	697.99	622.29
山　西 Shanxi	2845.96	1693.87	30.09	28.61	3.53	30.78	886.39	410.18
内 蒙 古 Inner Mongolia	1231.67	1041.07	2.36	2.35	0.12	28.30	720.44	707.11
辽　宁 Liaoning	940.40	658.14	6.99	5.84	1.20	24.32	291.77	269.74
吉　林 Jilin	939.89	716.00	6.88	6.83	0.01	27.98	252.43	245.25
黑 龙 江 Heilongjiang	1815.50	1630.32	3.18	2.59		26.19	377.93	370.92
上　海 Shanghai	15.73	14.00				66.34	10.44	9.63
江　苏 Jiangsu	621.37	526.46	9.59	9.35	1.70	45.06	147.54	138.94
浙　江 Zhejiang	2594.23	1928.95	75.28	64.33	28.88	46.05	431.25	390.29
安　徽 Anhui	3041.96	2527.15	95.69	77.44	12.48	34.78	724.13	631.63
福　建 Fujian	3461.20	2443.08	55.98	43.37	7.68	43.85	998.18	897.56
江　西 Jiangxi	6987.01	5626.43	201.76	188.78	16.43	36.59	1866.80	1377.70
山　东 Shandong	913.22	697.72	49.01	44.94	26.57	35.51	326.20	308.59
河　南 Henan	12755.57	10211.49	413.70	333.67	127.28	33.29	2381.89	2138.49
湖　北 Hubei	2521.11	1876.99	95.09	64.09	29.98	34.12	660.58	564.11
湖　南 Hunan	4271.88	3375.27	147.67	104.66	3.88	31.10	883.28	729.10
广　东 Guangdong	94.77	55.83	7.73	4.60	3.18	32.14	13.12	12.47
广　西 Guangxi	2543.09	2055.39	62.32	55.91	7.29	32.07	686.48	650.66
海　南 Hainan	63.01	57.73	0.27	0.27		30.12	16.31	16.31
重　庆 Chongqing	1000.01	868.93	8.66	8.43	0.03	34.72	208.43	201.36
四　川 Sichuan	2425.76	1949.60	45.30	39.95	0.68	36.13	593.83	541.79
贵　州 Guizhou	7667.09	5718.03	65.55	58.90	2.42	78.63	505.69	436.43
云　南 Yunnan	5137.85	3828.32	120.82	101.73	3.46	37.09	1317.84	1194.37
西　藏 Tibet	2121.18	507.72	25.40	20.94	1.45	63.75	420.60	258.53
陕　西 Shaanxi	128.36	105.16	16.69	9.65		33.94	29.16	27.58
甘　肃 Gansu	1289.33	974.65	23.86	22.35	0.95	28.98	574.91	529.33
青　海 Qinghai	867.99	679.86	7.88	7.83		31.11	231.04	220.20
宁　夏 Ningxia	515.42	420.30	8.53	8.38	0.14	27.66	225.26	200.80
新　疆 Xinjiang	3806.83	2473.90	117.97	111.35	0.83	37.88	4033.19	2014.71

Building Construction of Townships(2022)

Public Building			生产性建筑　Industrial Building					地区名称
本年竣工建筑面积（万平方米）Floor Space Completed This Year (10,000 m²)	混合结构以上 Mixed Strucure and Above	房地产开发 Real Estate Development	年末实有建筑面积（万平方米）Total Floor Space of Buildings (year-end) (10,000 m²)	混合结构以上 Mixed Strucure and Above	本年竣工建筑面积（万平方米）Floor Space Completed This Year (10,000 m²)	混合结构以上 Mixed Strucure and Above	房地产开发 Real Estate Development	Name of Regions
711.71	613.75	35.17	12232.41	9358.13	534.00	448.56	21.89	全　国
			7.66	7.66				北　京
			7.64	6.65				天　津
20.39	12.67	1.20	775.20	677.98	17.90	15.85	2.27	河　北
11.69	11.67		585.26	364.48	19.42	19.37		山　西
3.88	3.72		242.97	223.55	3.38	1.65		内　蒙　古
0.66	0.66		136.45	106.95				辽　宁
15.95	15.95		261.73	230.71	2.91	2.84		吉　林
2.59	2.59	1.00	301.16	284.61	2.36	2.36	1.00	黑　龙　江
			8.55	8.09				上　海
1.03	1.03		298.12	272.00	13.58	13.58	0.60	江　苏
69.82	69.40		357.47	305.94	59.33	56.44		浙　江
49.30	34.21	3.73	513.62	414.03	24.35	23.60	0.01	安　徽
16.36	15.39	0.10	650.33	500.95	18.63	18.02		福　建
90.91	86.11	2.40	1159.62	908.78	55.66	46.24	1.50	江　西
8.88	7.88	2.57	328.89	278.91	14.37	14.37	0.20	山　东
116.11	91.60	13.52	2087.35	1730.22	71.24	50.27	4.27	河　南
37.28	31.38	1.65	477.48	369.01	34.25	29.86	2.60	湖　北
47.76	45.03	2.61	635.28	425.88	46.43	27.29	2.09	湖　南
0.66	0.66		4.85	3.79	0.40	0.40		广　东
13.95	13.49	2.00	468.81	277.27	21.32	21.18		广　西
0.05	0.05		11.01	10.90	0.64	0.64		海　南
4.17	4.17		80.19	66.49	2.15	2.03		重　庆
27.73	26.96	0.02	397.22	307.68	10.67	8.05	0.20	四　川
37.08	20.41	0.23	274.68	188.97	21.79	13.02	0.20	贵　州
38.73	33.38	0.18	878.30	551.04	32.14	26.62		云　南
36.57	31.99		298.39	46.91	2.30	2.01		西　藏
1.70	1.70		11.08	9.33	0.16	0.16		陕　西
7.94	7.73		365.80	295.82	4.61	4.51		甘　肃
4.24	3.72		41.05	37.70	1.71	1.70		青　海
3.35	1.82	0.35	196.21	154.32	8.07	6.54	0.10	宁　夏
42.96	38.39	3.61	370.03	291.51	44.22	39.95	6.85	新　疆

3-2-16 乡建设投入(2022年)

计量单位: 万元

地区名称 Name of Regions	本年建设投入合计 Total Construction Input of This Year	房屋 Building					小计 Input of Municipal Public Facilities	供水 Water Supply
		小计 Input of House	房地产开发 Real Estate Development	住宅 Residential Building	公共建筑 Public Building	生产性建筑 Industrial Building		
全 国 National Total	4811648	3373241	304431	2027678	778485	567079	1438407	149449
北 京 Beijing	6972	2773	360	2773			4198	340
天 津 Tianjin	6						6	
河 北 Hebei	237263	197747	7168	154373	24753	18622	39516	3368
山 西 Shanxi	131930	75630	4790	55116	10621	9893	56299	12014
内 蒙 古 Inner Mongolia	30537	17232	320	4262	8986	3984	13306	1481
辽 宁 Liaoning	13690	7111	901	4667	2263	181	6579	490
吉 林 Jilin	49268	31130	236	9610	17338	4182	18137	820
黑 龙 江 Heilongjiang	23773	12135	4	3852	5721	2562	11638	231
上 海 Shanghai	399						399	
江 苏 Jiangsu	63695	44987	3650	21669	4533	18785	18708	734
浙 江 Zhejiang	241404	148149	11010	81673	45036	21440	93255	11898
安 徽 Anhui	324758	232431	39866	109367	44797	78267	92327	11819
福 建 Fujian	213222	125478	29030	82718	23653	19107	87744	8251
江 西 Jiangxi	555781	373993	55528	239198	94364	40431	181788	13761
山 东 Shandong	105012	91475	30055	58728	10262	22485	13538	2381
河 南 Henan	814719	609179	66235	411196	129957	68026	205540	13208
湖 北 Hubei	243716	146166	13949	70894	51822	23450	97550	7565
湖 南 Hunan	277608	202944	5960	126525	46280	30139	74664	9338
广 东 Guangdong	19507	13532		10810	2122	600	5975	545
广 西 Guangxi	162877	124363	1953	62143	20762	41458	38514	4332
海 南 Hainan	4476	2720	32	798	22	1900	1756	12
重 庆 Chongqing	42852	15562	462	8652	4823	2086	27291	2286
四 川 Sichuan	126670	93443	3025	45580	39024	8839	33227	4374
贵 州 Guizhou	164623	121667	5588	59751	23303	38613	42955	8463
云 南 Yunnan	362278	255976	5742	165481	52079	38415	106303	16331
西 藏 Tibet	120987	88771	1563	59997	25369	3405	32216	3384
陕 西 Shaanxi	13338	10974		9236	1626	112	2364	165
甘 肃 Gansu	83773	49657	1520	29202	15000	5455	34116	3623
青 海 Qinghai	39090	22668		11201	8259	3208	16422	726
宁 夏 Ningxia	30433	19958	1952	7765	5124	7070	10475	1195
新 疆 Xinjiang	306990	235389	13531	120440	60585	54364	71602	6313

Construction Input of Townships(2022)

Measurement Unit: 10,000 RMB

市政公用设施　Municipal Public Facilities									地区名称
燃　气	集中供热	道路桥梁	排　水	污水处理	园林绿化	环境卫生	垃圾处理	其　他	
Gas Supply	Central Heating	Road and Bridge	Drainage	Wastewater Treatment	Landscaping	Environmental Sanitation	Garbage Treatment	Other	Name of Regions
31433	25879	433265	326369	218202	134450	212918	114547	124645	全　国
	150	350	580	510	401	1674	515	703	北　京
						6	6		天　津
3723	1416	12749	4296	1030	6965	5129	2345	1870	河　北
659	6690	2646	12956	2236	4788	16243	1531	304	山　西
60	1262	4033	1939	822	957	2572	1584	1002	内　蒙古
15	1714	1587	473	44	400	1803	1212	98	辽　宁
141	1196	6674	5169	3826	709	2443	1594	984	吉　林
8	323	4505	1261	1017	923	3285	2058	1102	黑龙江
		150	29	2		120	70	100	上　海
1061		2406	3142	2827	2730	3902	1533	4736	江　苏
466	0	19348	25590	11124	11140	12428	6050	12384	浙　江
2658	15	19703	28450	21303	6995	12537	7315	10150	安　徽
63		30704	23385	18664	7941	14284	9914	3117	福　建
995	82	64986	40903	26664	15429	24441	13039	21191	江　西
700	400	3475	1695	727	1942	2374	919	572	山　东
12665	3308	82901	26575	15865	26365	30415	15286	10103	河　南
1214		46983	17506	12986	7361	9272	6810	7649	湖　北
929	26	13365	21912	17241	5231	16450	8711	7413	湖　南
		2223	2555	2555	420	227	155	5	广　东
190		14173	7948	4472	4391	4881	3477	2599	广　西
		140	57	51	645	901	30		海　南
186		11798	5866	5254	1588	2388	1341	3179	重　庆
2111		8603	9664	7253	1973	5413	3883	1090	四　川
197	128	13350	11789	9025	1068	6261	4573	1700	贵　州
559		21495	36944	29264	7215	13371	9284	10387	云　南
	2660	9035	3889	1441	1507	1196	628	10544	西　藏
60	1	440	197	47	181	641	218	679	陕　西
122	571	13654	6769	3906	2249	5362	3646	1767	甘　肃
		6975	2938	2530	1194	3520	1581	1068	青　海
36	759	2368	2826	1481	588	2241	943	462	宁　夏
2618	5180	12445	19067	14039	11152	7139	4295	7688	新　疆

3-2-17　镇乡级特殊区域市政公用设施水平(2022年)

地区名称 Name of Regions	人口密度 (人/平方公里) Population Density (person/km²)	人均日生 活用水量 (升) Per Capita Daily Water Consumption (liter)	供 水 普及率 (%) Water Coverage Rate (%)	燃 气 普及率 (%) Gas Coverage Rate (%)	人均道 路面积 (平方米) Road Surface Area Per Capita (m²)	排水管道 暗渠密度 (公里/平方公里) Density of Drains (km/km²)
全　国　National Total	**3050**	**115.85**	**91.84**	**65.76**	**25.95**	**7.63**
北　京　Beijing	3464	197.75	100.00	39.81	8.14	4.85
河　北　Hebei	2826	94.70	88.96	76.27	24.58	7.01
山　西　Shanxi	2680	102.68	92.08	29.45	72.52	2.99
内 蒙 古　Inner Mongolia	1878	89.71	82.33	4.52	62.42	3.58
辽　宁　Liaoning	1987	88.63	36.22	62.41	20.96	1.56
吉　林　Jilin	4091	88.76	99.12	26.18	23.96	7.21
黑 龙 江　Heilongjiang	441	83.33	76.23	17.55	80.16	1.46
上　海　Shanghai	2002	86.23	71.28	27.61	6.35	2.06
江　苏　Jiangsu	4469	93.80	97.67	82.54	17.18	10.54
浙　江　Zhejiang	1346	82.88	100.00	84.03	75.63	8.37
安　徽　Anhui	6505	87.47	77.17	58.77	18.59	18.28
福　建　Fujian	5947	118.02	76.67	75.34	21.19	11.49
江　西　Jiangxi	2946	104.04	95.80	67.41	21.00	6.68
山　东　Shandong	1793	81.39	99.08	69.23	22.93	6.43
河　南　Henan	4119	63.64	98.17	78.22	23.65	16.30
湖　北　Hubei	2647	103.73	92.75	54.96	35.02	10.00
湖　南　Hunan	4710	111.63	66.67	26.16	23.25	15.40
广　东　Guangdong	3350	196.48	100.00	100.00	20.99	9.54
广　西　Guangxi	1127	122.48	100.00	81.27	28.14	3.55
海　南　Hainan	8031	103.68	99.64	92.41	9.77	11.06
四　川　Sichuan	2191	51.89	87.76	94.62	26.39	9.48
云　南　Yunnan	4608	100.35	86.77	17.65	28.93	8.91
甘　肃　Gansu	7939	124.53	50.38		30.73	9.24
宁　夏　Ningxia	3747	98.25	96.01	57.76	21.35	6.61
新　疆　Xinjiang	2601	96.92	89.62	40.31	44.71	31.53
新疆生产 建设兵团　Xinjiang Production and Construction Corps	3419	127.82	98.25	69.17	26.71	7.68

Level of Municipal Public Facilities of Built-up Area of Special District at Township Level(2022)

污水处理率 (%) Wastewater Treatment Rate (%)	污水处理厂集中处理率 Centralized Treatment Rate of Wastewater Treatment Plants (%)	人均公园绿地面积(平方米) Public Recreational Green Space Per Capita (m²)	绿化覆盖率 (%) Green Coverage Rate (%)	绿地率 (%) Green Space Rate (%)	生活垃圾处理率 (%) Domestic Garbage Treatment Rate (%)	无害化处理率 Domestic Garbage Harmless Treatment Rate (%)	地区名称 Name of Regions
56.61	51.13	4.02	20.90	15.59	92.17	60.43	全　国
			98.20	98.20	100.00	100.00	北　京
61.64	42.82	2.77	12.02	6.42	100.00	81.99	河　北
		1.12	16.35	7.08	23.53	20.29	山　西
1.03		0.03	16.17	8.85	27.30	16.97	内 蒙 古
16.72	16.72	4.17	6.39	5.19	83.08	74.22	辽　宁
		1.60	13.37	10.18	100.00	90.89	吉　林
		6.41	15.27	5.27	28.71	27.70	黑 龙 江
88.37	88.37	10.01	28.68	28.57	98.70	98.70	上　海
78.05	60.59	9.78	24.38	19.64	99.38	57.00	江　苏
26.32	1.32		10.41	6.11	78.26	78.26	浙　江
9.72	9.72	2.28	26.86	12.63	99.70	99.70	安　徽
92.89	75.70	2.78	23.04	19.34	100.00	100.00	福　建
49.08	27.94	2.20	9.68	6.58	91.72	52.07	江　西
28.50	25.46	6.65	17.23	13.68	100.00	100.00	山　东
15.19	10.64	32.37	36.30	15.70	65.98	44.42	河　南
70.78	46.91	0.92	22.20	13.12	99.08	95.15	湖　北
		9.75	18.01	14.04	67.09	30.96	湖　南
3.96		0.25	16.26	5.55	100.00	100.00	广　东
95.57	95.57		23.78	10.22	100.00	89.78	广　西
86.63	86.63	3.57	13.39	7.78	99.16	21.85	海　南
70.79	49.84	0.55	3.29	1.79	100.00	100.00	四　川
2.60	2.60	1.19	7.98	5.63	52.31	16.73	云　南
100.00	100.00		36.36	32.58	100.00	100.00	甘　肃
31.07	18.93	1.69	17.20	13.29	84.05	69.26	宁　夏
21.02	17.66	0.89	19.67	10.05	85.60	42.92	新　疆
65.25	63.03	4.45	26.11	21.52	95.26	56.13	新疆生产建设兵团

3-2-18　镇乡级特殊区域基本情况(2022年)

地区名称 Name of Regions	镇乡级特殊 区域个数 (个) Number of Special District at Township Level (unit)	建成区 面积 (公顷) Surface Area of Built-up Districts (hectare)	建成区 户籍人口 (万人) Registered Permanent Population (10,000 persons)	建成区 常住人口 (万人) Permanant Population (10,000 persons)
全　国　National Total	407	60583.37	148.09	184.77
北　京　Beijing	1	132.00	0.39	0.46
河　北　Hebei	18	1451.59	4.10	4.10
山　西　Shanxi	4	100.33	0.27	0.27
内 蒙 古　Inner Mongolia	30	1958.70	3.79	3.68
辽　宁　Liaoning	24	4713.72	8.64	9.37
吉　林　Jilin	6	166.30	0.75	0.68
黑 龙 江　Heilongjiang	39	3858.72	3.16	1.70
上　海　Shanghai	1	1447.00	0.79	2.90
江　苏　Jiangsu	8	967.57	3.94	4.32
浙　江　Zhejiang	1	442.03	1.43	0.60
安　徽　Anhui	13	765.95	4.45	4.98
福　建　Fujian	8	756.00	2.66	4.50
江　西　Jiangxi	26	2099.97	8.56	6.19
山　东　Shandong	5	2769.00	5.76	4.97
河　南　Henan	3	135.00	0.36	0.56
湖　北　Hubei	26	3131.00	8.04	8.29
湖　南　Hunan	16	446.28	1.57	2.10
广　东　Guangdong	6	589.14	2.28	1.97
广　西　Guangxi	3	1148.63	0.78	1.29
海　南　Hainan	6	1728.81	2.60	13.88
四　川　Sichuan	1	173.00	0.33	0.38
云　南　Yunnan	15	604.40	2.64	2.79
甘　肃　Gansu	1	6.60	0.06	0.05
宁　夏　Ningxia	14	1056.00	4.34	3.96
新　疆　Xinjiang	34	1895.66	4.58	4.93
新疆生产 建设兵团　Xinjiang Production and Construction Corps	98	28039.97	71.83	95.87

Summary of Special District at Township Level(2022)

设有村镇建设管理机构的个数（个）Number of Towns with Construction Management Institution (unit)	村镇建设管理人员（人）Number of Construction Management Personnel (person)	规划建设管理　Planning and Administer		有总体规划的镇乡级特殊区域个数（个）Number of Special District at Township Level with Master Plans (unit)	本年编制 Compiled This Year	本年规划编制投入（万元）Input in Planning This Year (10,000 RMB)	地区名称 Name of Regions
		专职人员 Full-time Staff					
317	2476	1300		274	12	1532.70	全　国
1	6			1			北　京
16	37	31		12			河　北
3	3	2					山　西
26	93	72		20	1		内　蒙　古
23	50	37		15			辽　宁
4	6	6		2			吉　林
17	24	13		5			黑　龙　江
1				1			上　海
4	19	7		7			江　苏
				1			浙　江
9	173	102		8			安　徽
6	12	7		8		58.20	福　建
24	124	48		22	1	166.00	江　西
5	18	16		4		17.00	山　东
3	10	6		3			河　南
24	169	82		21	2	167.00	湖　北
10	19	7		7	1	15.50	湖　南
4	17	6		2			广　东
3	126	50		1			广　西
4	7	2		3			海　南
1	6	2		1		20.00	四　川
12	148	91		8			云　南
1	2	1		1			甘　肃
13	54	38		11		12.50	宁　夏
23	42	27		25		118.00	新　疆
80	1311	647		85	7	958.50	新疆生产建设兵团

3-2-19 镇乡级特殊区域供水(2022年)

地区名称 Name of Regions	集中供水的镇乡级特殊区域个数(个) Number of Special District at Township Level with Access to Piped Water (unit)	占全部镇乡级特殊区域的比例(%) Percentage of Total Rate (%)	公共供水综合生产能力(万立方米/日) Integrated Production Capacity of Public Water Supply Facilities (10,000 m³/day)	自备水源单位综合生产能力(万立方米/日) Integrated Production Capacity of Self-built Water Supply Facilities (10,000 m³/day)	年供水总量(万立方米) Annual Supply of Water (10,000 m³)
全 国 National Total	376	92.38	91.15	26.44	13716.15
北 京 Beijing	1	100.00	2.00	2.00	46.50
河 北 Hebei	18	100.00	2.94	0.55	422.82
山 西 Shanxi	4	100.00	0.10	0.01	14.38
内 蒙 古 Inner Mongolia	29	96.67	1.82	0.68	233.89
辽 宁 Liaoning	21	87.50	6.95	5.05	192.85
吉 林 Jilin	6	100.00	0.51	0.10	28.28
黑 龙 江 Heilongjiang	17	43.59	0.77	0.07	53.41
上 海 Shanghai	1	100.00	0.55		125.00
江 苏 Jiangsu	8	100.00	2.09	0.29	308.43
浙 江 Zhejiang	1	100.00	1.40		225.00
安 徽 Anhui	13	100.00	1.59	0.27	178.97
福 建 Fujian	8	100.00	1.26	0.08	290.75
江 西 Jiangxi	26	100.00	2.47	1.10	414.94
山 东 Shandong	5	100.00	5.25	0.20	1075.91
河 南 Henan	3	100.00	0.15	0.14	47.18
湖 北 Hubei	26	100.00	5.51	1.66	553.31
湖 南 Hunan	16	100.00	1.72	1.13	123.02
广 东 Guangdong	6	100.00	1.63	0.14	260.10
广 西 Guangxi	3	100.00	2.33	0.50	186.48
海 南 Hainan	6	100.00	1.74	1.59	972.75
四 川 Sichuan	1	100.00	0.30	0.30	44.33
云 南 Yunnan	14	93.33	1.44	1.05	354.55
甘 肃 Gansu	1	100.00	0.01		1.20
宁 夏 Ningxia	14	100.00	4.07	0.71	529.52
新 疆 Xinjiang	31	91.18	1.35	0.39	245.24
新疆生产建设兵团 Xinjiang Production and Construction Corps	97	98.98	41.20	8.46	6787.34

Water Supply of Special District at Township Level(2022)

年生活 用水量 Annual Domestic Water Consumption	年生产 用水量 Annual Water Consumption for Production	供水管道 长　度 （公里） Length of Water Supply Pipelines (km)	本年新增 Added This Year	用水人口 （万人） Population with Access to Water (10,000 persons)	地区名称 Name of Regions
7175.39	5772.07	12020.71	382.03	169.69	全　国
33.00	13.50	31.45	0.65	0.46	北　京
126.16	252.06	193.41		3.65	河　北
9.28	4.90	67.40		0.25	山　西
99.18	59.86	285.00	16.00	3.03	内　蒙　古
109.75	63.79	164.10		3.39	辽　宁
21.85	4.83	39.20		0.67	吉　林
39.43	13.31	167.87	0.40	1.30	黑　龙　江
65.00	60.00	100.00		2.07	上　海
144.60	161.38	208.63		4.22	江　苏
18.00	200.00	40.00		0.60	浙　江
122.77	52.46	267.22	3.00	3.85	安　徽
148.50	131.68	166.30	10.00	3.45	福　建
225.11	150.00	238.14	3.40	5.93	江　西
146.15	922.46	151.40	3.00	4.92	山　东
12.68	26.00	96.00	1.00	0.55	河　南
291.01	240.46	873.44	38.46	7.69	湖　北
57.09	36.02	222.18	10.10	1.40	湖　南
141.52	118.58	184.30	18.00	1.97	广　东
57.85	128.63	48.20	8.00	1.29	广　西
523.55	365.22	271.40	3.17	13.83	海　南
6.30	7.00	81.00	16.00	0.33	四　川
88.52	261.89	275.96		2.42	云　南
1.20		2.25		0.03	甘　肃
136.21	388.16	266.60	3.81	3.80	宁　夏
156.33	79.60	624.12	15.85	4.42	新　疆
4394.35	2030.28	6955.14	231.19	94.19	新疆生产 建设兵团

3-2-20 镇乡级特殊区域燃气、供热、道路桥梁(2022年)

地区名称 Name of Regions	用气人口 (万人) Population with Access to Gas (10,000 persons)	集中供热 (万平方米) Area of Centrally Heated District (10,000 m^2)	道路长度 (公里) Length of Roads (km)	本年新增 Added This Year	本年更新 改　造 Renewal and Upgrading during The Reported Year
全　国　**National Total**	**121.50**	**4403.73**	**6707.56**	**177.88**	**155.64**
北　京　Beijing	0.18	21.00	5.37		
河　北　Hebei	3.13	35.30	185.42	1.16	6.74
山　西　Shanxi	0.08	3.20	45.35		0.20
内 蒙 古　Inner Mongolia	0.17	16.96	482.69	2.00	2.10
辽　宁　Liaoning	5.85	335.28	260.78		20.17
吉　林　Jilin	0.18	1.35	31.40		
黑 龙 江　Heilongjiang	0.30	24.30	231.40		1.20
上　海　Shanghai	0.80		21.58		
江　苏　Jiangsu	3.57		112.26		
浙　江　Zhejiang	0.50		48.00		2.00
安　徽　Anhui	2.93		209.44	4.20	6.11
福　建　Fujian	3.39		121.53	4.55	10.30
江　西　Jiangxi	4.17		257.13	27.49	18.70
山　东　Shandong	3.44	48.20	122.86	10.26	1.30
河　南　Henan	0.44		29.70		
湖　北　Hubei	4.55		447.37	22.15	14.12
湖　南　Hunan	0.55		116.24	16.20	8.62
广　东　Guangdong	1.97		77.28	1.94	1.94
广　西　Guangxi	1.05		71.14		0.43
海　南　Hainan	12.83		165.27		
四　川　Sichuan	0.36		25.00	7.00	7.00
云　南　Yunnan	0.49		164.69	0.91	2.00
甘　肃　Gansu			2.00		
宁　夏　Ningxia	2.29	85.58	156.97	0.22	4.62
新　疆　Xinjiang	1.99	102.11	379.11		0.30
新疆生产 建设兵团　Xinjiang Production and Construction Corps	66.31	3730.45	2937.58	79.80	47.79

Gas, Central Heating, Road and Bridge of Special District at Township Level(2022)

安装路灯的道路长度 Roads with Lamps	道路面积（万平方米）Surface Area of Roads (10,000 m²)	本年新增 Added This year	本年更新改造 Renewal and Upgrading during The Reported Year	桥梁座数（座）Number of Bridges (unit)	本年新增 Added This year	地区名称 Name of Regions
1533.41	4795.54	129.16	104.09	952	17	全 国
5.37	3.72					北 京
71.35	100.85	1.04	3.05	25		河 北
6.10	19.50		0.60	6		山 西
40.76	229.66	0.90	4.05	39	2	内 蒙 古
99.69	196.36		8.40	47		辽 宁
10.50	16.30					吉 林
24.53	136.32		0.74	8		黑 龙 江
	18.39			4	1	上 海
33.93	74.30			28		江 苏
	45.00	5.00		32		浙 江
17.75	92.62	1.77	0.98	28	1	安 徽
34.70	95.26	4.90	11.95	17		福 建
67.55	129.91	8.80	7.98	25		江 西
69.37	113.87	5.68	4.12	24		山 东
2.00	13.15			4		河 南
125.60	290.22	19.16	3.74	44	5	湖 北
13.88	48.87	8.71	1.25	20	2	湖 南
26.00	41.42	1.16	1.36	12		广 东
3.30	36.41		0.20	3		广 西
15.65	135.69			9		海 南
9.00	10.00	2.80	2.80	5	2	四 川
37.70	80.57	1.16	2.20	15		云 南
2.00	1.61					甘 肃
38.19	84.46	0.96	2.09	25		宁 夏
99.15	220.48	0.25	7.19	25	1	新 疆
679.34	2560.60	66.87	41.39	507	3	新疆生产建设兵团

3-2-21 镇乡级特殊区域排水和污水处理(2022年)
Drainage and Wastewater Treatment of Special District at Township Level(2022)

地区名称 Name of Regions	对生活污水进行处理的镇乡级特殊区域 Special District at Township Level with		污水处理厂 Wastewater Treatment Plant		污水处理装置处理能力(万立方米/日) Treatment Capacity of Wastewater Treatment Facilities (10,000 m³/day)	排水管道长度(公里) Length of Drainage Pipelines (km)	本年新增 Added This Year	排水暗渠长度(公里) Length of Drains (km)	本年新增 Added This Year
	个数(个) Number (unit)	比例(%) Rate (%)	个数(个) Number of Wastewater Treatment Plants (unit)	处理能力(万立方米/日) Treatment Capacity (10,000 m³/day)					
全 国 National Total	233	57.25	142	24.35	25.66	3320.52	94.82	1300.85	14.84
北 京 Beijing						6.40			
河 北 Hebei	11	61.11	4	1.18	0.47	75.00	0.50	26.79	
山 西 Shanxi	1	25.00				1.60		1.40	
内 蒙 古 Inner Mongolia	3	10.00		0.00	0.01	49.73		20.40	
辽 宁 Liaoning	6	25.00	3	0.14	0.14	45.01	0.02	28.60	1.00
吉 林 Jilin	2	33.33				3.00		8.99	
黑 龙 江 Heilongjiang	1	2.56				39.76		16.73	
上 海 Shanghai	1	100.00				29.77			
江 苏 Jiangsu	7	87.50	3	0.55	0.81	79.60	1.00	22.36	1.10
浙 江 Zhejiang	1	100.00				37.00			
安 徽 Anhui	4	30.77			0.06	56.35	0.21	83.69	0.71
福 建 Fujian	8	100.00	2	1.85	3.11	77.35	5.50	9.53	
江 西 Jiangxi	18	69.23	15	0.22	1.03	70.56	4.70	69.75	2.40
山 东 Shandong	5	100.00	4	1.30	0.48	123.60	3.50	54.50	0.50
河 南 Henan	2	66.67	1	0.01	0.02	12.00		10.00	
湖 北 Hubei	26	100.00	15	0.85	0.86	172.95	3.67	140.01	2.32
湖 南 Hunan	2	12.50				18.58	2.11	50.14	1.61
广 东 Guangdong	1	16.67			0.10	24.30		31.90	
广 西 Guangxi	1	33.33	1	1.00	1.00	24.68		16.05	
海 南 Hainan	4	66.67			3.25	0.00	130.13	7.40	61.10
四 川 Sichuan	1	100.00			0.32	0.32	16.20	0.20	0.20	0.20
云 南 Yunnan	3	20.00		0.00	0.05	34.64	10.90	19.20	5.00
甘 肃 Gansu	1	100.00	1	0.01	0.01	0.60		0.01	
宁 夏 Ningxia	11	78.57	9	0.43	0.40	63.55	2.40	6.20	
新 疆 Xinjiang	17	50.00	2	0.17	0.16	95.44	0.60	502.30	
新疆生产建设兵团 Xinjiang Production and Construction Corps	96	97.96	82	13.08	16.64	2032.72	52.11	121.00	

3-2-22 镇乡级特殊区域园林绿化及环境卫生(2022年)
Landscaping and Environmental Sanitation of Special District at Township Level(2022)

地区名称 Name of Regions			园林绿化(公顷) Landscaping((hectare)				环境卫生 Environmental Sanitation		
			绿化覆盖面积 Green Coverage Area	绿地面积 Area of Parks and Green Space	本年新增 Added This Year	公园绿地面积 Public Green Space	生活垃圾中转站 (座) Number of Garbage Transfer Station (unit)	环卫专用车辆设备 (辆) Number of Special Vehicles for Environmental Sanitation (unit)	公共厕所 (座) Number of Latrines (unit)
全	国	National Total	12663.00	9442.67	436.54	742.38	335	1372	1736
北	京	Beijing	129.62	129.62	1.60		5	9	5
河	北	Hebei	174.54	93.12	2.24	11.35	8	48	68
山	西	Shanxi	16.40	7.10	1.70	0.30	12	28	6
内 蒙 古		Inner Mongolia	316.63	173.39	1.11	0.12	39	107	166
辽	宁	Liaoning	301.14	244.51	4.27	39.05	6	109	61
吉	林	Jilin	22.23	16.93	1.00	1.09		12	13
黑 龙 江		Heilongjiang	589.31	203.42	4.80	10.90		45	87
上	海	Shanghai	415.00	413.34	0.33	29.00	1	8	5
江	苏	Jiangsu	235.93	190.04	1.18	42.29	8	31	29
浙	江	Zhejiang	46.00	27.00	1.00				
安	徽	Anhui	205.71	96.76	1.82	11.37	25	30	75
福	建	Fujian	174.22	146.19	2.10	12.51	7	24	25
江	西	Jiangxi	203.21	138.23	5.40	13.61	29	41	100
山	东	Shandong	477.13	378.93	7.00	33.00	7	31	20
河	南	Henan	49.00	21.20	0.20	18.00	4	9	7
湖	北	Hubei	695.13	410.71	5.90	7.63	19	94	170
湖	南	Hunan	80.38	62.68	1.60	20.50	10	23	37
广	东	Guangdong	95.80	32.70	0.50	0.50		20	22
广	西	Guangxi	273.12	117.41	0.58		1	6	7
海	南	Hainan	231.51	134.44	62.04	49.55	4	23	17
四	川	Sichuan	5.70	3.10	2.20	0.21	5	8	8
云	南	Yunnan	48.26	34.02	1.51	3.32	1	17	106
甘	肃	Gansu	2.40	2.15	0.20			1	5
宁	夏	Ningxia	181.58	140.29	0.79	6.70	10	46	59
新	疆	Xinjiang	372.94	190.56	15.06	4.40	30	74	109
新疆生产建设兵团		Xinjiang Production and Construction Corps	7320.11	6034.83	310.41	426.98	100	528	529

3-2-23　镇乡级特殊区域房屋(2022年)

地区名称 Name of Regions	住宅　Residential Building						公共建筑	
	年末实有建筑面积(万平方米) Total Floor Space of Buildings (year-end) (10,000 m²)	混合结构以上 Mixed Strucure and Above	本年竣工建筑面积(万平方米) Floor Space Completed This Year (10,000 m²)	混合结构以上 Mixed Strucure and Above	房地产开发 Real Estate Development	人均住宅建筑面积(平方米) Per Capita Floor Space (m²)	年末实有建筑面积(万平方米) Total Floor Space of Buildings (year-end) (10,000 m²)	混合结构以上 Mixed Strucure and Above
全　国　National Total	**7247.03**	**5634.81**	**146.16**	**116.74**	**76.20**	**48.94**	**1737.49**	**1524.68**
北　京　Beijing	31.47	31.47				79.75		
河　北　Hebei	200.85	171.85				49.02	26.69	25.95
山　西　Shanxi	6.49	1.89	0.04	0.01	0.03	23.70	1.23	
内 蒙 古　Inner Mongolia	97.52	72.61	0.58	0.58		25.74	24.10	23.22
辽　宁　Liaoning	440.99	420.33				51.05	44.63	43.23
吉　林　Jilin	29.66	20.29				39.59	2.12	2.08
黑 龙 江　Heilongjiang	95.15	93.12				30.08	19.43	18.55
上　海　Shanghai	139.00	139.00				176.98		
江　苏　Jiangsu	110.38	84.78	4.30	4.30		28.04	30.61	19.01
浙　江　Zhejiang	16.00	15.00				11.19	1.10	1.10
安　徽　Anhui	215.98	188.30				48.49	74.78	74.64
福　建　Fujian	177.75	154.27	7.78	7.54	6.84	66.87	22.64	21.47
江　西　Jiangxi	232.05	211.23	11.22	11.12		27.11	26.50	24.68
山　东　Shandong	223.51	168.84	5.69	5.50	3.19	38.78	104.39	98.93
河　南　Henan	13.66	13.59				38.30	2.22	2.22
湖　北　Hubei	356.10	330.13	11.57	9.72	1.81	44.31	85.12	83.57
湖　南　Hunan	76.80	56.60	2.21	2.11		49.05	4.92	4.91
广　东　Guangdong	105.46	94.21	0.47	0.47		46.28	8.26	6.01
广　西　Guangxi	27.78	25.60				35.69	5.17	5.02
海　南　Hainan	99.86	94.86	6.83	6.83	6.45	38.43	19.57	15.05
四　川　Sichuan	19.70	19.01	0.34	0.19	0.15	58.98	4.35	4.29
云　南　Yunnan	90.12	76.82	3.34	1.72		34.19	81.80	73.60
甘　肃　Gansu	0.80	0.80				13.56	1.00	1.00
宁　夏　Ningxia	172.47	160.64				39.70	33.12	27.78
新　疆　Xinjiang	170.65	80.68	0.27	0.24		37.22	39.23	30.74
新疆生产建设兵团　Xinjiang Production and Construction Corps	4096.84	2908.90	91.52	66.40	57.73	57.04	1074.52	917.63

Building Construction of Special District at Township Level(2022)

Public Building			生产性建筑 Industrial Building					地区名称
本年竣工建筑面积（万平方米）Floor Space Completed This Year (10,000 m²)	混合结构以上 Mixed Strucure and Above	房地产开发 Real Estate Development	年末实有建筑面积（万平方米）Total Floor Space of Buildings (year-end) (10,000 m²)	混合结构以上 Mixed Strucure and Above	本年竣工建筑面积（万平方米）Floor Space Completed This Year (10,000 m²)	混合结构以上 Mixed Strucure and Above	房地产开发 Real Estate Development	Name of Regions
37.67	33.70	4.15	2338.19	1843.88	88.15	59.45	0.15	全 国
								北 京
0.08	0.08		44.58	38.45				河 北
			1.02					山 西
0.83	0.83		16.38	14.11	1.00	1.00		内 蒙 古
			270.63	265.30				辽 宁
			1.05	0.97				吉 林
0.07	0.07		43.86	22.96				黑 龙 江
			35.50	35.50				上 海
			70.61	62.51	3.00	3.00		江 苏
0.10	0.10		86.00	86.00	2.90			浙 江
			29.10	28.49				安 徽
1.37	1.37		249.74	249.21	20.02	20.02		福 建
1.01	0.90		30.00	24.33	0.32	0.32		江 西
2.06	2.06	0.20	425.39	350.26	20.40	12.30		山 东
			1.28	1.20				河 南
7.01	6.99		91.79	78.76	1.40	1.40		湖 北
0.23	0.23		6.17	5.08	0.15	0.15		湖 南
			5.05	5.05	1.24	1.24		广 东
			26.35	17.63				广 西
0.65	0.56		24.96	22.52	17.58	1.80		海 南
0.38	0.19	0.19	2.91	2.84	0.29	0.11	0.15	四 川
0.20	0.20		12.53	11.57	1.16	0.17		云 南
			0.15	0.15				甘 肃
0.01	0.01		127.81	101.68				宁 夏
0.15	0.15		21.45	14.71				新 疆
23.54	19.98	3.76	713.90	404.59	18.69	17.94		新疆生产建设兵团

3-2-24 镇乡级特殊区域建设投入(2022年)

计量单位: 万元

地区名称 Name of Regions	合计 Total Construction Input of This Year	房屋 Building					小计 Input of Municipal Public Facilities	供水 Water Supply
		本年建设投入小计 Input of House	房地产开发 Real Estate Development	住宅 Residential Building	公共建筑 Public Building	生产性建筑 Industrial Building		
全 国 National Total	831066	624131	142791	308270	113341	202520	206935	27443
北 京 Beijing	719						719	110
河 北 Hebei	1714	183			183		1531	96
山 西 Shanxi	24	23		23			1	
内 蒙 古 Inner Mongolia	4763	3695		920	1860	915	1068	
辽 宁 Liaoning	2138	47		47			2091	30
吉 林 Jilin	15						15	
黑 龙 江 Heilongjiang	287	93			93		194	5
上 海 Shanghai	6550	1000	1000	1000			5550	150
江 苏 Jiangsu	21269	20000		12000		8000	1269	63
浙 江 Zhejiang	950	900			400	500	50	
安 徽 Anhui	11150	10800				10800	350	34
福 建 Fujian	62085	51990	16150	17350	3300	31340	10095	57
江 西 Jiangxi	86812	74599		70898	2926	775	12213	290
山 东 Shandong	58359	46797	11962	12806	3640	30351	11562	219
河 南 Henan								
湖 北 Hubei	24723	16100		8285	6161	1654	8622	1069
湖 南 Hunan	4710	3756		1678	1210	868	954	211
广 东 Guangdong	11190	10647		137	210	10300	543	
广 西 Guangxi	243						243	74
海 南 Hainan	51270	49569	15000	15502	2267	31800	1701	
四 川 Sichuan	2548	1066		351	394	321	1482	169
云 南 Yunnan	7263	5486		4273	240	973	1777	16
甘 肃 Gansu	31						31	
宁 夏 Ningxia	1210	50			50		1160	78
新 疆 Xinjiang	4245	751		351	400		3494	486
新疆生产建设兵团 Xinjiang Production and Construction Corps	466800	326579	98679	162649	90007	73923	140221	24287

Construction Input of Special District at Township Level(2022)

Measurement Unit: 10,000 RMB

燃 气 Gas Supply	集中供热 Central Heating	道路桥梁 Road and Bridge	排 水 Drainage	污水处理 Wastewater Treatment	园林绿化 Landscaping	环境卫生 Environmental Sanitation	垃圾处理 Garbage Treatment	其 他 Other	地区名称 Name of Regions
5412	35824	28659	40968	30050	17858	25450	9122	25320	全 国
					513	96	36		北 京
40	332	165	147	13	266	403	162	84	河 北
						1	1		山 西
	50	810			29	173	147	6	内 蒙 古
	148	503	249	217	136	643	388	383	辽 宁
						3	12	8	吉 林
						19	170		黑 龙 江
2600		700	600		400	1100	120		上 海
18		181	286	260	169	280	85	272	江 苏
		30			20				浙 江
36		94	42		35	109	25		安 徽
20		4924	3810	3679	400	780	420	105	福 建
0		1163	8380	7934	823	916	291	641	江 西
112	62	2410	718	604	2643	4717	3732	681	山 东
									河 南
502		2276	2742	2655	814	943	447	276	湖 北
12		155	142	119	161	222	109	51	湖 南
						543	125		广 东
		51	55		8	55	30		广 西
			1510	1510	15	176	156		海 南
		230	410	410	413	260	230		四 川
		650	80		679	193	120	159	云 南
						31	31		甘 肃
		225	546	546	107	179	103	27	宁 夏
1246	2	907	5	5	271	576	403	2	新 疆
826	35231	13187	21247	12099	9934	12874	1953	22633	新疆生产 建设兵团

3-2-25　村庄人口及面积(2022年)

地区名称 Name of Regions	村庄建设用地面积 (公顷) Area of Villages Construction Land (hectare)	行政村个数(个) Number of Administrative Villages(unit)		
		合　计 Total	500人以下 Under 500 (persons)	500-1000人 500-1000 (persons)
全　国　National Total	12490687.91	477874	59643	106198
北　京　Beijing	88655.44	3473	1026	1077
天　津　Tianjin	63495.69	2924	777	1065
河　北　Hebei	850188.61	44495	10144	14740
山　西　Shanxi	312602.91	18582	3819	6705
内　蒙　古　Inner Mongolia	263446.66	10992	2526	3607
辽　宁　Liaoning	407098.95	10512	253	1401
吉　林　Jilin	333568.31	9150	1330	2179
黑　龙　江　Heilongjiang	437568.16	8921	1152	1681
上　海　Shanghai	65925.00	1510	119	148
江　苏　Jiangsu	666915.88	13565	158	584
浙　江　Zhejiang	303376.53	15948	912	4127
安　徽　Anhui	607139.27	14495	312	921
福　建　Fujian	269450.85	13471	1385	3278
江　西　Jiangxi	445330.14	16805	974	2881
山　东　Shandong	1022144.25	56673	15477	18870
河　南　Henan	1002893.42	42106	2487	8507
湖　北　Hubei	459386.63	20715	1479	4827
湖　南　Hunan	675809.12	23346	601	3325
广　东　Guangdong	601969.19	18066	781	1890
广　西　Guangxi	475147.55	14220	174	1109
海　南　Hainan	107759.81	2775	160	388
重　庆　Chongqing	221742.58	8300	246	853
四　川　Sichuan	747362.76	26251	2473	3023
贵　州　Guizhou	345840.00	13431	571	1993
云　南　Yunnan	493876.20	13316	208	1148
西　藏　Tibet	45360.13	5209	3356	1417
陕　西　Shaanxi	337347.09	15941	1152	4057
甘　肃　Gansu	480414.97	15882	2080	5602
青　海　Qinghai	55797.20	4096	1295	1634
宁　夏　Ningxia	67025.13	2231	139	365
新　疆　Xinjiang	208713.62	8771	1059	2286
新疆生产 建设兵团　Xinjiang Production and Construction Corps	27335.86	1702	1018	510

Population and Area of Villages(2022)

1000人以上 Above 1000 (persons)	自然村个数 （个） Number of Natural Villages (unit)	村庄户籍 人 口 （万人） Registered Permanent Population (10,000 persons)	村庄常住 人 口 （万人） Permanart Population (10,000 persons)	地区名称 Name of Regions
312033	2332112	77221.79	63567.87	全 国
1370	4586	330.46	462.83	北 京
1082	2947	234.11	219.51	天 津
19611	66488	4683.01	4054.20	河 北
8058	41966	1909.83	1575.45	山 西
4859	45907	1339.56	961.20	内 蒙 古
8858	49155	1712.83	1465.45	辽 宁
5641	39367	1302.80	995.53	吉 林
6088	34623	1628.74	1168.42	黑 龙 江
1243	17769	308.36	457.07	上 海
12823	122457	3390.16	3228.35	江 苏
10909	74069	2054.78	2018.90	浙 江
13262	191673	4442.88	3501.50	安 徽
8808	63761	2003.83	1594.37	福 建
12950	155444	3107.04	2474.65	江 西
22326	86928	5269.61	4707.38	山 东
31112	184583	6765.85	5666.50	河 南
14409	103810	3310.36	2633.69	湖 北
19420	106467	4282.56	3403.43	湖 南
15395	145798	4678.38	3726.85	广 东
12937	167979	3983.52	2870.66	广 西
2227	18691	572.05	523.25	海 南
7201	56952	1890.64	1128.33	重 庆
20755	121479	5701.60	4123.12	四 川
10867	72391	2751.86	2160.06	贵 州
11960	129438	3382.97	3100.55	云 南
436	19609	245.64	233.40	西 藏
10732	69124	2157.69	1874.47	陕 西
8200	86101	1863.85	1555.37	甘 肃
1167	11571	355.17	335.33	青 海
1727	12745	370.31	267.21	宁 夏
5426	26400	1099.13	1021.34	新 疆
174	1834	92.19	59.52	新疆生产 建设兵团

3-2-26 村庄公共设施(一)(2022年)

地区名称 Name of Regions	集中供水的行政村 Administrative Villages With Access to Piped Water		年生活用 水量 (万立方米) Annual Domestic Water Consumption (10,000 m³)	供水管 道长度 (公里) Length of Water Supply Pipelines (km)	本年新增 Added This Year
	个数 (个) Number (unit)	比例 (%) Rate (%)			
全 国 National Total	366772	84.85	1974281.51	2194211.41	82138.85
北 京 Beijing	2584	82.08	19090.66	16292.14	265.21
天 津 Tianjin	2321	95.59	8150.33	14448.76	142.57
河 北 Hebei	35108	86.65	119903.02	193711.89	3669.00
山 西 Shanxi	14553	84.11	42754.40	65736.35	1297.50
内 蒙 古 Inner Mongolia	7591	74.18	22895.04	51999.69	720.70
辽 宁 Liaoning	6198	65.45	38006.12	44442.27	1003.03
吉 林 Jilin	7585	91.68	26993.48	65982.12	2356.16
黑 龙 江 Heilongjiang	7349	93.99	32036.63	65906.61	314.45
上 海 Shanghai	1261	93.75	17230.40	9698.26	19.34
江 苏 Jiangsu	12233	99.00	109773.85	105446.88	2055.43
浙 江 Zhejiang	11220	84.41	99110.02	70591.39	3197.42
安 徽 Anhui	11478	88.86	98835.24	92473.79	4355.82
福 建 Fujian	10580	91.63	55531.42	43011.86	1909.91
江 西 Jiangxi	11284	73.56	62475.63	56500.78	3465.67
山 东 Shandong	48607	96.23	153556.56	169966.78	4402.93
河 南 Henan	33904	88.37	152895.51	128318.51	4305.80
湖 北 Hubei	15934	82.82	109960.79	93489.51	4121.42
湖 南 Hunan	14447	67.63	80115.37	79639.32	3331.06
广 东 Guangdong	14603	87.57	134223.29	101525.72	10518.91
广 西 Guangxi	9853	74.22	93486.85	66856.62	2833.80
海 南 Hainan	2464	93.94	18161.34	12641.35	441.87
重 庆 Chongqing	6774	87.58	36272.07	42625.50	2397.19
四 川 Sichuan	18703	79.31	119275.75	126179.23	6388.19
贵 州 Guizhou	9167	76.88	64717.40	82655.45	4306.09
云 南 Yunnan	9952	85.10	98962.33	118739.14	6441.46
西 藏 Tibet	2188	44.69	10174.11	12602.59	289.15
陕 西 Shaanxi	13146	92.92	54301.14	60499.12	2234.65
甘 肃 Gansu	12107	82.84	40950.39	81961.02	2941.04
青 海 Qinghai	3085	79.59	9889.91	22048.51	981.68
宁 夏 Ningxia	1909	99.43	8996.36	19370.20	433.32
新 疆 Xinjiang	7283	88.86	33363.95	69634.35	593.85
新疆生产建设兵团 Xinjiang Production and Construction Corps	1301	78.94	2192.15	9215.70	404.23

Public Facilities of Villages Ⅰ (2022)

用水人口（万人）Population with Access to Water (10,000 persons)	供水普及率（%）Water Coverage Rate (%)	人均日生活用水量（升）Per Capita Daily Water Consumption (liter)	用气人口（万人）Population with Access to Gas (10,000 persons)	燃气普及率（%）Gas Coverage Rate (%)	集中供热面积（万平方米）Area of Centrally Heated District (10,000 m²)	地区名称 Name of Regions
54683.11	86.02	98.92	25380.64	39.93	38130.35	全　国
437.54	94.54	119.54	247.80	53.54	1403.96	北　京
212.03	96.59	105.32	180.96	82.44	1611.86	天　津
3799.88	93.73	86.45	2902.60	71.59	7322.95	河　北
1388.70	88.15	84.35	328.60	20.86	5949.27	山　西
715.12	74.40	87.71	129.44	13.47	1156.90	内　蒙　古
1062.79	72.52	97.97	256.81	17.52	1397.60	辽　宁
833.09	83.68	88.77	136.36	13.70	654.91	吉　林
1015.55	86.92	86.43	67.36	5.77	603.63	黑　龙　江
428.04	93.65	110.29	358.54	78.44	133.21	上　海
3175.39	98.36	94.71	2824.66	87.50	11.65	江　苏
1778.92	88.11	152.64	1034.44	51.24	202.75	浙　江
2852.08	81.45	94.94	1482.37	42.34		安　徽
1475.43	92.54	103.12	988.27	61.99		福　建
1745.77	70.55	98.05	777.24	31.41	612.56	江　西
4527.15	96.17	92.93	2719.97	57.78	11485.22	山　东
4781.22	84.38	87.61	1550.78	27.37	1139.72	河　南
2134.49	81.05	141.14	902.58	34.27	281.30	湖　北
2317.89	68.10	94.70	966.21	28.39	37.01	湖　南
3435.02	92.17	107.05	2645.26	70.98		广　东
2470.45	86.06	103.68	1738.26	60.55	124.18	广　西
481.28	91.98	103.38	388.74	74.29		海　南
991.33	87.86	100.24	374.14	33.16		重　庆
3209.04	77.83	101.83	1514.17	36.72		四　川
1832.32	84.83	96.77	112.69	5.22	5.63	贵　州
2743.09	88.47	98.84	154.01	4.97		云　南
201.65	86.40	138.23	28.65	12.27	24.70	西　藏
1665.10	88.83	89.35	338.97	18.08	689.13	陕　西
1381.12	88.80	81.23	93.74	6.03	1291.84	甘　肃
323.23	96.39	83.83	15.43	4.60	222.50	青　海
258.79	96.85	95.24	31.74	11.88	434.71	宁　夏
960.23	94.02	95.19	73.69	7.21	1212.27	新　疆
49.36	82.93	121.67	16.16	27.16	120.89	新疆生产建设兵团

3-2-27 村庄公共设施(二)(2022年)

地区名称 Name of Regions	村庄内 道路长度 (公里) The Length of Roads within Villages (kilometer)	本年新增 Added This Year	本年更新 改造 Renewal and Upgrading during The Reported Year	硬化道路 Hardened Roads
全 国 National Total	**3573922.53**	**69089.67**	**82160.11**	**2244118.83**
北 京 Beijing	19381.84	285.27	1817.88	15808.70
天 津 Tianjin	16431.62	21.18	676.03	14018.30
河 北 Hebei	250654.42	1768.09	4862.12	186602.34
山 西 Shanxi	86068.36	419.06	949.16	49823.82
内 蒙 古 Inner Mongolia	85215.09	675.63	724.03	56727.06
辽 宁 Liaoning	78467.65	1548.33	2843.46	49063.41
吉 林 Jilin	86619.82	780.86	2448.54	66058.94
黑 龙 江 Heilongjiang	82037.01	210.59	1079.85	49621.14
上 海 Shanghai	11453.59	19.33	374.45	8770.72
江 苏 Jiangsu	142715.55	2193.00	2887.14	111847.77
浙 江 Zhejiang	84214.74	1414.17	2447.50	40124.65
安 徽 Anhui	170173.85	4344.10	3203.78	111188.04
福 建 Fujian	75817.23	1210.67	1328.33	47259.31
江 西 Jiangxi	103116.53	3092.44	3267.29	59103.15
山 东 Shandong	350582.42	4099.02	9714.93	251668.11
河 南 Henan	209248.73	5431.87	4776.63	122542.65
湖 北 Hubei	215526.53	4900.84	6223.43	95283.32
湖 南 Hunan	177922.57	3202.16	4402.47	86609.03
广 东 Guangdong	174620.00	4901.17	4657.55	99917.63
广 西 Guangxi	120843.64	2004.82	3280.00	87728.52
海 南 Hainan	30513.17	519.44	688.97	8514.46
重 庆 Chongqing	39719.81	1556.07	919.98	26664.40
四 川 Sichuan	305436.16	8701.07	6186.79	215995.96
贵 州 Guizhou	133642.81	2782.44	1271.26	60850.38
云 南 Yunnan	164261.01	4247.70	2976.61	96868.00
西 藏 Tibet	17680.08	177.43	290.31	5987.14
陕 西 Shaanxi	105140.47	2562.22	2564.37	78916.63
甘 肃 Gansu	102700.52	3368.98	2934.28	64963.05
青 海 Qinghai	29989.32	325.68	457.68	15366.49
宁 夏 Ningxia	27456.94	741.68	703.61	20925.25
新 疆 Xinjiang	67265.99	1472.45	980.40	35749.31
新疆生产 建设兵团 Xinjiang Production and Construction Corps	9005.06	111.91	221.28	3551.15

Public Facilities of Villages Ⅱ (2022)

村庄内道路面积（万平方米）The Area of Roads within Villages (10,000 m²)	本年新增 Added This Year	本年更新改造 Renewal and Upgrading during The Reported Year	硬化道路 Hardened Roads	排水管道沟渠长度（公里）The Length of Drainage Pipelines and Canals (kilometer)	本年新增 Added This Year	地区名称 Name of Regions
2546996.31	70456.71	66791.70	1265920.02	1280372.23	38964.76	全　国
11078.63	192.89	961.85	9033.52	13462.45	333.69	北　京
7624.58	13.86	378.95	6135.84	6911.67	189.94	天　津
119250.67	2196.04	2870.15	84436.58	60242.99	679.51	河　北
68802.96	504.06	795.77	32873.47	54968.53	530.90	山　西
46832.58	586.27	726.54	28500.44	10022.06	236.07	内　蒙　古
44136.54	1149.63	2059.28	25114.30	33217.72	273.51	辽　宁
42362.91	432.79	1490.22	30569.73	46878.93	821.16	吉　林
41111.67	111.42	586.14	21777.64	34136.01	165.18	黑　龙　江
6837.36	15.47	220.63	5447.80	6996.67	11.87	上　海
135219.75	1461.50	2244.71	85778.73	53899.70	2481.07	江　苏
71196.89	1627.62	2225.43	29781.25	43567.81	1017.24	浙　江
151327.30	5136.54	2570.99	57903.36	50030.89	2254.05	安　徽
44263.69	1038.79	1024.59	24057.03	61003.79	1111.58	福　建
81554.92	2967.12	2837.43	34480.11	41871.70	2175.44	江　西
209992.69	4368.05	8259.91	137274.36	230347.28	5459.61	山　东
201038.34	6066.12	4327.96	79919.91	59684.23	2691.30	河　南
201122.33	5191.72	6739.60	56558.78	62412.84	1739.64	湖　北
146441.57	3378.61	3884.74	47951.80	52513.85	1280.95	湖　南
139568.30	6082.19	4449.08	62717.99	68737.81	3630.72	广　东
71392.76	1890.24	1858.55	46067.26	31084.27	790.30	广　西
16089.75	616.55	409.64	4636.53	5261.54	290.74	海　南
23414.68	1085.52	512.11	15511.72	15937.75	553.88	重　庆
206979.27	8428.98	4975.58	111637.68	80186.98	2304.94	四　川
127457.77	5806.53	1248.60	58146.60	24913.77	1008.78	贵　州
107588.92	3399.74	3424.97	51324.01	46364.07	2756.68	云　南
12984.60	329.04	277.53	4139.68	2733.92	99.46	西　藏
56650.51	1899.81	1855.09	34595.56	34013.80	1342.82	陕　西
54721.18	2079.85	1844.35	32292.90	23013.37	1271.42	甘　肃
14192.89	182.39	274.99	7425.26	5163.67	110.68	青　海
19809.83	802.41	541.48	13635.17	11317.03	269.41	宁　夏
57266.67	1240.45	768.42	22029.90	7477.97	960.58	新　疆
8683.80	174.51	146.42	4165.11	1997.16	121.64	新疆生产建设兵团

3-2-28 村庄房屋(2022年)

地区名称 Name of Regions	住宅 Residential Building						公共建筑	
	年末实有建筑面积 (万平方米) Total Floor Space of Buildings (year-end) (10,000 m²)	混合结构以上 Mixed Strucure and Above	本年竣工建筑面积 (万平方米) Floor Space Completed This Year (10,000 m²)	混合结构以上 Mixed Strucure and Above	房地产开发 Real Estate Development	人均住宅建筑面积 (平方米) Per Capita Floor Space (m²)	年末实有建筑面积 (万平方米) Total Floor Space of Buildings (year-end) (10,000 m²)	混合结构以上 Mixed Strucure and Above
全 国 National Total	2698071.74	2090936.17	43848.04	36326.33	4196.04	34.94	208943.29	165685.54
北 京 Beijing	16755.53	14907.60	304.62	258.17	52.32	50.70	3647.46	3535.04
天 津 Tianjin	7705.30	6554.07	44.01	41.82	16.00	32.91	522.03	509.37
河 北 Hebei	135994.36	108781.09	1042.60	839.34	114.14	29.04	4512.68	4131.14
山 西 Shanxi	56526.50	37581.74	1636.09	442.95	82.49	29.60	9016.73	5932.56
内 蒙 古 Inner Mongolia	36268.07	32313.00	57.40	53.34	5.95	27.07	2373.31	2161.71
辽 宁 Liaoning	39842.21	28035.33	222.46	154.33	19.32	23.26	2905.82	2306.29
吉 林 Jilin	32428.84	25157.21	81.60	56.12	20.71	24.89	1303.47	1210.55
黑 龙 江 Heilongjiang	39620.39	35464.79	113.67	108.60	28.30	24.33	1585.50	1555.51
上 海 Shanghai	12526.28	11206.18	60.17	41.57	18.60	40.62	919.85	698.83
江 苏 Jiangsu	146526.42	126557.80	1495.09	1219.02	269.64	43.22	14442.41	8764.21
浙 江 Zhejiang	99054.59	82319.79	1798.46	1598.22	332.53	48.21	11513.78	9344.66
安 徽 Anhui	152183.40	120652.53	2005.57	1695.02	135.16	34.25	7986.43	5660.51
福 建 Fujian	75929.03	57839.76	1319.94	1191.30	245.56	37.89	7438.03	6025.87
江 西 Jiangxi	128460.29	107279.48	2148.26	1961.90	34.17	41.34	8485.25	7287.32
山 东 Shandong	170775.08	130658.37	2297.55	2012.24	577.77	32.41	16062.68	13672.05
河 南 Henan	223220.86	184075.69	5997.60	5549.98	307.49	32.99	13837.48	12813.00
湖 北 Hubei	111677.11	87819.47	2284.54	1780.16	106.68	33.74	8203.85	7030.54
湖 南 Hunan	194678.74	160744.54	3925.21	2978.56	95.43	45.46	15860.40	13497.97
广 东 Guangdong	130294.22	101503.15	4447.56	3212.44	1318.03	27.85	18610.58	14730.30
广 西 Guangxi	124650.96	98533.26	1519.34	1462.17	11.45	31.29	13630.77	12319.35
海 南 Hainan	17645.50	16705.65	329.63	287.88	20.61	30.85	785.36	676.17
重 庆 Chongqing	75660.00	61699.95	668.72	638.54	11.69	40.02	2827.02	2675.25
四 川 Sichuan	214109.04	171537.48	2776.91	2235.72	176.45	37.55	8488.33	7631.53
贵 州 Guizhou	82981.10	60846.41	2736.28	2561.24	35.40	30.15	5395.01	3398.98
云 南 Yunnan	129765.29	88511.11	2221.73	2052.45	79.34	38.36	7235.66	6215.93
西 藏 Tibet	12028.07	2917.78	102.91	57.52		48.97	5368.36	656.77
陕 西 Shaanxi	115360.29	53881.96	642.71	560.54	17.71	53.46	3826.45	3510.78
甘 肃 Gansu	54043.05	39183.25	838.27	652.99	21.71	29.00	7083.33	3968.61
青 海 Qinghai	10947.59	7178.12	55.31	52.72		30.82	528.93	434.41
宁 夏 Ningxia	11417.72	8845.59	132.83	119.51	2.06	30.83	702.57	661.55
新 疆 Xinjiang	37123.32	20694.48	497.69	430.14	25.36	33.78	2984.77	2443.73
新疆生产建设兵团 Xinjiang Production and Construction Corps	1872.60	949.56	43.33	19.84	13.97	20.31	859.01	225.10

Building Construction of Villages(2022)

本年竣工建筑面积（万平方米）Floor Space Completed This Year (10,000 m²)	混合结构以上 Mixed Strucure and Above	房地产开发 Real Estate Development	年末实有建筑面积（万平方米）Total Floor Space of Buildings (year-end) (10,000 m²)	混合结构以上 Mixed Strucure and Above	本年竣工建筑面积（万平方米）Floor Space Completed This Year (10,000 m²)	混合结构以上 Mixed Strucure and Above	房地产开发 Real Estate Development	地区名称 Name of Regions
5059.09	4160.14	201.67	331808.29	262884.13	9703.88	8253.91	442.09	全 国
2.71	0.45	2.16	5931.94	5483.70				北 京
3.73	3.08	0.39	2090.43	1810.26	20.14	19.79	0.20	天 津
85.20	76.49	1.97	8099.02	6791.15	180.04	139.59	3.29	河 北
53.86	40.19	0.40	11053.14	5321.25	205.20	173.36	1.10	山 西
13.98	12.53		4824.69	2708.80	160.06	150.81		内 蒙 古
78.06	75.54		6809.19	3729.67	50.08	28.45	0.64	辽 宁
37.95	37.03	0.10	1945.35	1742.62	49.61	48.29	7.28	吉 林
19.48	18.68	1.00	3849.68	3573.49	21.68	20.10	1.05	黑 龙 江
14.63	14.43		4622.86	4388.28	25.26	25.16		上 海
195.83	190.41	26.57	37924.61	30965.73	787.98	706.81	18.34	江 苏
248.33	212.02	4.90	22850.54	20377.72	644.51	589.61	8.41	浙 江
287.71	250.43	16.73	12209.30	9109.92	338.58	287.87	29.64	安 徽
180.15	158.10	11.24	16364.71	11893.01	357.33	289.15	119.09	福 建
602.31	469.82	3.76	7389.03	6153.09	466.87	338.33	7.82	江 西
523.09	385.84	51.05	37383.18	30872.85	731.17	578.28	79.34	山 东
286.00	253.00	17.75	24038.05	18851.48	338.41	288.79	22.58	河 南
389.46	275.56	22.78	9109.12	6942.52	365.84	316.03	29.33	湖 北
631.69	431.57	8.52	10016.83	8307.21	501.39	438.55	34.70	湖 南
298.20	252.18	5.44	41483.90	37606.53	1143.92	1002.75	26.43	广 东
123.13	111.93	1.84	7985.52	5922.22	327.23	186.95	0.36	广 西
15.88	14.34	0.81	390.00	333.00	17.93	8.06	0.87	海 南
34.43	33.68	5.06	3646.51	2902.40	65.50	55.97	0.17	重 庆
195.55	176.71	4.47	11746.07	9421.89	339.74	250.08	20.14	四 川
174.26	161.57	4.44	5598.51	3348.22	264.96	215.73	1.26	贵 州
188.87	160.02	2.83	12047.93	8872.39	1592.56	1475.19	27.42	云 南
22.28	17.43		1124.60	651.66	6.94	5.14	0.04	西 藏
152.88	142.33	3.00	4451.00	3498.37	229.62	175.11	0.37	陕 西
89.64	85.14	1.09	7246.31	5480.25	183.20	173.13	0.20	甘 肃
7.61	7.45		578.07	394.76	28.88	28.32		青 海
6.47	5.03		1391.36	1081.97	75.10	69.08		宁 夏
92.98	84.44	3.36	6922.53	3992.98	148.44	143.66	2.04	新 疆
2.76	2.73		684.33	354.77	35.73	25.80		新疆生产建设兵团

3-2-29　村庄建设投入(2022年)

计量单位: 万元

地区名称 Name of Regions	本年建设投入合计 Total Construction Input of This Year	房屋　Building					小　计 Input of Municipal Public Facilities	供　水 Water Supply
		小　计 Input of House	房地产开发 Real Estate Development	住　宅 Residential Building	公共建筑 Public Building	生产性建筑 Industrial Building		
全　国　National Total	**88494217**	**61889589**	**10480562**	**44380979**	**6151891**	**11356719**	**26604628**	**3181879**
北　京　Beijing	988323	480445	109277	455316	25068	60	507879	59506
天　津　Tianjin	164170	77081	3572	50402	4799	21881	87089	3707
河　北　Hebei	2192469	1453660	134716	1128886	110967	213806	738808	94875
山　西　Shanxi	1914039	1492797	77478	1280848	117655	94294	421242	46340
内 蒙 古　Inner Mongolia	294165	136348	1957	65237	11199	59912	157817	21620
辽　宁　Liaoning	497084	228686	9411	120514	19480	88692	268397	19779
吉　林　Jilin	792972	356503	178167	204272	54855	97375	436469	47713
黑 龙 江　Heilongjiang	541747	382349	8922	288472	61108	32769	159398	10185
上　海　Shanghai	721121	256090	22101	109794	55137	91158	465032	8576
江　苏　Jiangsu	5682361	3881794	592635	2406892	302882	1172020	1800567	105737
浙　江　Zhejiang	6380337	4410982	1130795	2827374	379866	1203743	1969355	228882
安　徽　Anhui	4320355	2929160	268894	2126290	429158	373712	1391195	187991
福　建　Fujian	3897950	2996758	945222	2116756	284924	595078	901192	99285
江　西　Jiangxi	3606757	2421982	71354	1858666	317121	246195	1184774	132496
山　东　Shandong	8978899	6187710	2202195	4217950	827946	1141813	2791189	322355
河　南　Henan	6110239	4646891	151307	3956392	353965	336535	1463348	126596
湖　北　Hubei	3743237	2452039	227934	1766213	303072	382754	1291198	157183
湖　南　Hunan	4288585	3215440	266465	2376075	472617	366748	1073145	171103
广　东　Guangdong	10169834	7990980	2683948	4930675	551497	2508808	2178854	320433
广　西　Guangxi	2731885	2134844	39461	1552148	158275	424422	597041	114405
海　南　Hainan	628017	423045	47763	377353	25940	19752	204972	22516
重　庆　Chongqing	1367569	835412	51117	729840	43962	61610	532157	57028
四　川　Sichuan	5371539	3683323	284150	2978710	268376	436236	1688216	212906
贵　州　Guizhou	2438909	1813419	225273	1421945	179692	211782	625490	124921
云　南　Yunnan	4770558	3506080	525793	2810680	264381	431019	1264479	195488
西　藏　Tibet	278499	195499	1219	92764	85801	16934	83001	11898
陕　西　Shaanxi	1773167	1044078	41785	684345	177909	181825	729089	81308
甘　肃　Gansu	1677345	1026584	57068	756887	80167	189530	650761	68281
青　海　Qinghai	265179	122954	1225	84104	13158	25692	142225	64130
宁　夏　Ningxia	465127	250854	8644	166086	13410	71358	214273	10844
新　疆　Xinjiang	1144696	716078	103802	388003	139046	189029	428618	30921
新疆生产 建设兵团　Xinjiang Production and Construction Corps	297084	139725	6913	51090	18456	70178	157359	22870

Construction Input of Villages(2022)

Measurement Unit: 10,000 RMB

		市政公用设施 Municipal Public Facilities							地区名称
燃 气 Gas Supply	集中供热 Central Heating	道路桥梁 Road and Bridge	排 水 Drainage	污水处理 Wastewater Treatment	园林绿化 Landscaping	环境卫生 Environmental Sanitation	垃圾处理 Garbage Treatment	其 他 Other	Name of Regions
1182372	273640	10093890	4914505	3292072	1650024	3523789	1836387	1784530	全　　国
2032	1356	109222	189365	154296	59453	77182	24751	9764	北　京
2065	1106	38541	19891	15489	3907	15698	8059	2174	天　津
141280	37712	229563	63842	29928	31722	114108	66980	25708	河　北
57342	79815	87355	46454	18730	26585	67825	28885	9526	山　西
502	12384	72137	9104	4213	8318	27539	14775	6214	内　蒙　古
1174	6223	153376	18605	10840	8501	52838	32206	7902	辽　宁
8209	3397	193370	94860	39895	13999	53081	29120	21839	吉　林
250	1130	80631	11449	4219	5733	36762	22301	13258	黑　龙　江
4422		96782	226255	178903	23085	75408	26058	30503	上　海
56875	385	524058	483054	359280	168521	327102	157165	134835	江　苏
38320	1379	682046	351288	249987	176837	275482	134975	215121	浙　江
30566	737	591734	197173	115628	93928	221354	115178	67711	安　徽
12533		329561	194264	144034	63271	145633	97461	56644	福　建
6921	1518	529314	154919	73671	65915	159353	83472	134338	江　西
279141	74788	836650	521768	378722	201901	403242	204796	151344	山　东
191213	4071	550749	208860	115207	116748	203098	96001	62014	河　南
32865	1713	562177	183564	94428	111581	147325	81685	94790	湖　北
20242	1211	474861	141166	77306	46275	145935	88718	72351	湖　南
15243	1300	717304	742252	585952	94574	223196	125385	64552	广　东
3855	16	257462	97385	43883	12732	78246	56541	32939	广　西
5146	5200	78325	52500	34133	7027	25254	10273	9003	海　南
33331	31	286170	43483	29057	17542	48002	25429	46570	重　庆
90065		905465	200605	123067	66766	136293	82566	76117	四　川
23923	93	287596	89398	60235	15934	57785	36613	25841	贵　州
4529	191	474821	236292	158769	67076	112580	63393	173502	云　南
320	157	40415	13984	5106	1189	2149	852	12888	西　藏
49733	7591	309087	87309	35933	46340	93843	38499	53878	陕　西
9067	14726	305553	62842	31261	32450	63188	36072	94654	甘　肃
3113	18	40467	16397	10169	4239	12438	7749	1423	青　海
2694	1817	94017	31663	22307	12107	37461	11996	23672	宁　夏
47999	13068	133888	108932	77173	27813	40685	21991	25312	新　疆
7401	510	21197	15581	10251	17951	43705	6442	28145	新疆生产 建设兵团

主要指标解释

Explanatory Notes on Main Indicators

主要指标解释

城市和县城部分

人口密度

指城区内的人口疏密程度。计算公式：

$$人口密度 = \frac{城区人口 + 城区暂住人口}{城区面积}$$

人均日生活用水量

指每一用水人口平均每天的生活用水量。计算公式：

$$人均日生活用水量 = \frac{居民家庭用水量 + 公共服务用水量 + 免费供水量中的生活用水量}{用水人口} \div 报告期日历天数 \times 1000升$$

供水普及率

指报告期末城区内用水人口与总人口的比率。计算公式：

$$供水普及率 = \frac{城区用水人口（含暂住人口）}{城区人口 + 城区暂住人口}$$

$$公共供水普及率 = \frac{城区公共用水人口（含暂住人口）}{城区人口 + 城区暂住人口} \times 100\%$$

燃气普及率

指报告期末城区内使用燃气的人口与总人口的比率。计算公式：

$$燃气普及率 = \frac{城区用气人口（含暂住人口）}{城区人口 + 城区暂住人口} \times 100\%$$

人均城市道路面积

指报告期末城区内平均每人拥有的道路面积。计算公式：

$$人均道路面积 = \frac{城区道路面积}{城区人口 + 城区暂住人口}$$

建成区路网密度

指报告期末建成区内道路分布的稀疏程度。计算公式：

$$建成区路网密度 = \frac{建成区道路长度}{建成区面积}$$

建成区排水管道密度

指报告期末建成区排水管道分布的疏密程度。计算公式：

$$建成区排水管道密度 = \frac{建成区排水管道长度}{建成区面积}$$

污水处理率

指报告期内污水处理总量与污水排放总量的比率。计算公式：

$$污水处理率 = \frac{污水处理总量}{污水排放总量} \times 100\%$$

污水处理厂集中处理率

指报告期内通过污水处理厂处理的污水量与污水排放总量的比率。计算公式：

$$污水处理厂集中处理率=\frac{污水处理厂处理的污水量}{污水排放总量}\times100\%$$

人均公园绿地面积

指报告期末城区内平均每人拥有的公园绿地面积。计算公式：

$$人均公园绿地面积=\frac{城区公园绿地面积}{城区人口+城区暂住人口}$$

建成区绿化覆盖率

指报告期末建成区内绿化覆盖面积与区域面积的比率。计算公式：

$$建成区绿化覆盖率=\frac{建成区绿化覆盖面积}{建成区面积}\times100\%$$

建成区绿地率

指报告期末建成区内绿地面积与建成区面积的比率。计算公式：

$$建成区绿地率=\frac{建成区绿地面积}{建成区面积}\times100\%$$

生活垃圾处理率

指报告期内生活垃圾处理量与生活垃圾产生量的比率。计算公式：

$$生活垃圾处理率=\frac{生活垃圾处理量}{生活垃圾产生量}\times100\%$$

生活垃圾无害化处理率

指报告期内生活垃圾无害化处理量与生活垃圾产生量的比率。计算公式：

$$生活垃圾无害化处理率=\frac{生活垃圾无害化处理量}{生活垃圾产生量}\times100\%$$

在统计时，由于生活垃圾产生量不易取得，可用清运量代替。"垃圾清运量"在审核时要与总人口（包括暂住人口）对应，一般城市人均日产生垃圾为 1kg 左右。

市区（县）面积

指城市（县）行政区域内的全部土地面积（包括水域面积）。地级以上城市行政区不包括市辖县（市）。按国务院批准的行政区划面积为准填报。

城区（县城）面积

指城市的城区和县城的面积。

设市城市城区包括：市本级（1）街道办事处所辖地域；（2）城市公共设施、居住设施和市政公用设施等连接到的其他镇（乡）地域；（3）常住人口在 3000 人以上独立的工矿区、开发区、科研单位、大专院校等特殊区域。

县城包括：（1）县政府驻地的镇（城关镇）或街道办事处地域；（2）县城公共设施、居住设施和市政设施等连接到的其他镇（乡）地域；（3）县域内常住人口在 3000 人以上独立的工矿区、开发区、科研单位、大专院校等特殊区域。

连接是指两个区域间可观察到的已建成或在建的公共设施、居住设施、市政设施和其他设施相连，中间没有被水域、农业用地、园地、林地、牧草地等非建设用地隔断。

对于组团式和散点式的城市，城区由多个分散的区域组成，或有个别区域远离主城区，应将这些分散的区域相加作为城区。

在统计时，以镇（乡）一级为最小统计划分单位，原则上不要打破镇（乡）的行政区划。

城区（县城）人口

指划定的城区（县城）范围的户籍人口数。按公安部门的统计为准填报。

暂住人口

指离开常住户口地的市区或乡、镇，到本市居住半年以上的人员。按市区、县和城区、县城分别统计，一般按公安部门的暂住人口统计为准填报。

建成区面积

指城区（县城）内实际已成片开发建设、市政公用设施和公共设施基本具备的区域。对核心城市，它包括集中连片的部分以及分散的若干个已经成片建设起来，市政公用设施和公共设施基本具备的地区；对一城多镇来说，它包括由几个连片开发建设起来的，市政公用设施和公共设施基本具备的地区组成。因此建成区范围，一般是指建成区外轮廓线所能包括的地区，也就是这个城市实际建设用地所达到的范围。

城市建设用地面积

指城市内的居住用地、公共管理与公共服务用地、商业服务业设施用地、工业用地、物流仓储用地、道路与交通设施用地、公用设施用地、绿地与广场用地。分别统计规划建设用地和现状建设用地。

规划建设用地面积指截至报告期末对应有关城市、县人民政府所在地镇的最新版依法批准的城市总体规划确定的城市（镇）用地面积。

现状建设用地面积指报告期末对应有关城市、县人民政府所在地镇的城市建设用地实际情况的面积。

城市维护建设资金

指用于城市维护和建设的资金，资金来源包括城市维护建设税、公用事业附加、中央和地方财政拨款、国内贷款、债券收入、利用外资、土地出让转让收入、资产置换收入、市政公用企事业单位自筹资金、国家和省规定收取的用于城市维护建设的行政事业性收费、集资收入以及其他收入。资金支出包括固定资产投资支出、维护支出和其他支出。

本年鉴中仅统计了用于城市维护和建设的财政性资金。

城市维护建设税

指依据《中华人民共和国城市维护建设税暂行条例》开征的一种地方性税种。现行的征收办法是以纳税人实际缴纳的增值税、消费税、营业税税额为计税依据，与增值税、消费税、营业税同时缴纳。根据纳税人所在地不同执行不同的纳税率：市区的税率为百分之七；县城、镇的税率为百分之五；不在市区、县城或镇的税率为百分之一。

城市公用事业附加

指在部分公用事业产品（服务）价外征收的用于城市维护建设的附加收入。包括工业用电、工业用水附加，公共汽车、电车、民用自来水、民用照明用电、电话、煤气、轮渡等附加。

固定资产投资

指建造和购置市政公用设施的经济活动，即市政公用设施固定资产再生产活动。市政公用设施固定资产再生产过程包括固定资产更新（局部更新和全部更新）、改建、扩建、新建等活动。新的企业财务会计制度规定，固定资产局部更新的大修理作为日常生产活动的一部分，发生的大修理费用直接在成本费用中列支。按照现行投资管理体制及有关部门的规定，凡属于养护、维护性质的工程，不纳入固定资产投资统计。对新建和对现有市政公用设施改造工程，应纳入固定资产统计。

本年新增固定资产

指在报告期已经完成建造和购置过程，并交付生产或使用单位的固定资产价值。包括已经建成投入生产或交付使用的工程投资和达到固定资产标准的设备、工具、器具的投资及有关应摊入的费用。属于增加固定资产价值的其他建设费用，应随同交付使用的工程一并计入新增固定资产。

新增生产能力（或效益）

指通过固定资产投资活动而增加的设计能力。计算新增生产能力（或效益）是以能独立发挥生产能力

（或效益）的工程为对象。当工程建成，经有关部门验收鉴定合格，正式移交投入生产，即应计算新增生产能力（或效益）。

综合生产能力

指按供水设施取水、净化、送水、出厂输水干管等环节设计能力计算的综合生产能力。包括在原设计能力的基础上，经挖、革、改增加的生产能力。计算时，以四个环节中最薄弱的环节为主确定能力。对于经过更新改造，按更新改造后新的设计能力填报。

供水管道长度

指从送水泵至各类用户引入管之间所有市政管道的长度。不包括新安装尚未使用、水厂内以及用户建筑物内的管道。在同一条街道埋设两条或两条以上管道时，应按每条管道的长度计算。

供水总量

指报告期供水企业（单位）供出的全部水量。包括有效供水量和漏损水量。

有效供水量指水厂将水供出厂外后，各类用户实际使用到的水量。包括售水量和免费供水量。

漏损水量

指在供水工程中由于管道及附属设施破损而造成的漏失水量，以及计量损失量和其他损失水量。计量损失量包括居民用户总分表差损失水量、非居用户表误差损失水量；其他损失水量是指未注册用户用水和用户拒查等管理因素导致的损失水量。

新水取用量

指取自任何水源被第一次利用的水量，包括自来水、地下水、地表水。新水量就一个城市来说，包括城市供水企业新水量和社会各单位的新水量。

其中：**工业新水取用量**指为使工业生产正常进行，保证生产过程对水的需要，而实际从各种水源引取的、为任何目的所用的新鲜水量，包括间接冷却水新水量、工艺水新水量、锅炉新水量及其他新水量。

用水重复利用量

指各用水单位在生产和生活中，循环利用的水量和直接或经过处理后回收再利用的水量之和。

其中：**工业用水重复利用量**指工业企业内部生活及生产用水中，循环利用的水量和直接或经过处理后回收再利用的水量之和。

节约用水量

指报告期新节水量，通过采用各项节水措施（如改进生产工艺、技术、生产设备、用水方式、换装节水器具、加强管理等）后，用水量和用水效益产生效果，而节约的水量。

人工煤气生产能力

指报告期末燃气生产厂制气、净化、输送等环节的综合生产能力，不包括备用设备能力。一般按设计能力计算，如果实际生产能力大于设计能力时，应按实际测定的生产能力计算。测定时应以制气、净化、输送三个环节中最薄弱的环节为主。

供气管道长度

指报告期末从气源厂压缩机的出口或门站出口至各类用户引入管之间的全部已经通气投入使用的管道长度。不包括煤气生产厂、输配站、液化气储存站、灌瓶站、储配站、气化站、混气站、供应站等厂（站）内，以及用户建筑物内的管道。

供气总量

指报告期燃气企业（单位）向用户供应的燃气数量。包括销售量和损失量。

汽车加气站

指专门为燃气机动车（船舶）提供压缩天然气、液化石油气等燃料加气服务的站点。应按不同气种分别统计。

供热能力

指供热企业（单位）向城市热用户输送热能的设计能力。

供热总量

指在报告期供热企业（单位）向城市热用户输送全部蒸汽和热水的总热量。

供热管道长度

指从各类热源到热用户建筑物接入口之间的全部蒸汽和热水的管道长度。不包括各类热源厂内部的管道长度。

其中：**一级管网**指由热源至热力站间的供热管道，**二级管网**指热力站至用户之间的供热管道。

城市道路

指城市供车辆、行人通行的，具备一定技术条件的道路、桥梁、隧道及其附属设施。城市道路由车行道和人行道等组成。在统计时只统计路面宽度在 3.5 米（含 3.5 米）以上的各种铺装道路，包括开放型工业区和住宅区道路在内。

道路长度

指道路长度和与道路相通的桥梁、隧道的长度，按车行道中心线计算。

道路面积

指道路面积和与道路相通的广场、桥梁、隧道的铺装面积（统计时，将车行道面积、人行道面积分别统计）。

人行道面积按道路两侧面积相加计算，包括步行街和广场，不含人车混行的道路。

桥梁

指为跨越天然或人工障碍物而修建的构筑物。包括跨河桥、立交桥、人行天桥以及人行地下通道等。

道路照明灯盏数

指在城市道路设置的各种照明用灯。一根电杆上有几盏即计算几盏。统计时，仅统计功能照明灯，不统计景观照明灯。

防洪堤长度

指实际修筑的防洪堤长度。统计时应按河道两岸的防洪堤相加计算长度，但如河岸一侧有数道防洪堤时，只计算最长一道的长度。

污水排放总量

指生活污水、工业废水的排放总量，包括从排水管道和排水沟（渠）排出的污水量。

（1）可按每条管道、沟（渠）排放口的实际观测的日平均流量与报告期日历日数的乘积计算。

（2）有排水测量设备的，可按实际测量值计算。

（3）如无观测值，也可按当地供水总量乘以污水排放系数确定。

城市分类污水排放系数

城市污水分类	污水排放系数
城市污水	0.7～0.8
城市综合生活污水	0.8～0.9
城市工业废水	0.7～0.9

排水管道长度

指所有市政排水总管、干管、支管、检查井及连接井进出口等长度之和。计算时应按单管计算，即在同一条街道上如有两条或两条以上并排的排水管道时，应按每条排水管道的长度相加计算。

其中：**污水管道**指专门排放污水的排水管道。

雨水管道指专门排放雨水的排水管道。

雨污合流管道指雨水、污水同时进入同一管道进行排水的排水管道。

污水处理量

指污水处理厂（或污水处理装置）实际处理的污水量。包括物理处理量、生物处理量和化学处理量。

其中**处理本城区（县城）外**，指污水处理厂作为区域设施，不仅处理本城区（县城）的污水，还处理本市（县）以外其他市、县或本市（县）其他乡村等的污水。这部分污水处理量单独统计，并在计算本市（县）的污水处理率时扣除。

干污泥年产生量

指全年污水处理厂在污水处理过程中干污泥的最终产生量。干污泥是指以干固体质量计的污泥量，含水率为0。如果产生的湿污泥的含水率为n%，那么干污泥产生量=湿污泥产生量×（1-n%）。

干污泥处置量

指报告期内将污泥达标处理处置的干污泥量。统计时按土地利用、建材利用、焚烧、填埋和其他分别填写。其中：

污泥土地利用 指处理达标后的污泥产物用于园林绿化、土地改良、林地、农用等场合的处置方式。

污泥建筑材料利用 指将污泥处理达标后的产物作为制砖、水泥熟料等建筑材料部分原料的处置方式。

污泥焚烧 指利用焚烧炉将污泥完全矿化为少量灰烬的处理处置方式，包括单独焚烧，以及与生活垃圾、热电厂等工业窑炉的协同焚烧。

污泥填埋 指采取工程措施将处理达标后的污泥产物进行堆、填、埋，置于受控制场地内的处置方式。

绿化覆盖面积

指城市中乔木、灌木、草坪等所有植被的垂直投影面积。包括城市各类绿地绿化种植垂直投影面积、屋顶绿化植物的垂直投影面积以及零星树木的垂直投影面积，乔木树冠下的灌木和草本植物以及灌木树冠下的草本植物垂直投影面积均不能重复计算。

绿地面积

指报告期末用作园林和绿化的各种绿地面积。包括公园绿地、防护绿地、广场用地、附属绿地和位于建成区范围内的区域绿地面积。

其中：**公园绿地**指向公众开放，以游憩为主要功能，兼具生态、景观、文教和应急避险等功能，有一定游憩和服务设施的绿地。

防护绿地 指用地独立，具有卫生、隔离、安全、生态防护功能，游人不宜进入的绿地。主要包括卫生隔离防护绿地、道路及铁路防护绿地、高压走廊防护绿地、公共设施防护绿地等。

广场用地 指以游憩、纪念、集会和避险等功能为主的城市公共活动场地。

附属绿地 指附属于各类城市建设用地（除"绿地与广场用地"）的绿化用地。包括居住用地、公共管理与公共服务设施用地、商业服务业设施用地、工业用地、物流仓储用地、道路和交通设施用地、公共设施用地等用地中的绿地。

区域绿地 指位于城市建设用地之外，具有城乡生态环境及自然资源和文化资源保护、游憩健身、安全防护隔离、物种保护、园林苗木生产等功能的绿地。

公园

指常年开放的供公众游览、观赏、休憩、开展科学、文化及休闲等活动，有较完善的设施和良好的绿化环境、景观优美的公园绿地。包括综合性公园、儿童公园、文物古迹公园、纪念性公园、风景名胜公园、动物园、植物园、带状公园等。不包括居住小区及小区以下的游园。统计时只统计市级和区级的综合公园、专类公园和带状公园。

其中：**门票免费公园**指对公众免费开放，不售门票的公园。

道路清扫保洁面积

指报告期末对城市道路和公共场所（主要包括城市行车道、人行道、车行隧道、人行过街地下通道、道路附属绿地、地铁站、高架路、人行过街天桥、立交桥、广场、停车场及其他设施等）进行清扫保洁的面积。一天清扫保洁多次的，按清扫保洁面积最大的一次计算。

其中：**机械化道路清扫保洁面积**指报告期末使用扫路车（机）、冲洗车等大小型机械清扫保洁的道路面积。多种机械在一条道路上重复使用时，只按一种机械清扫保洁的面积计算，不能重复统计。

生活垃圾、建筑垃圾清运量

指报告期收集和运送到各生活垃圾、建筑垃圾厂和生活垃圾、建筑垃圾最终消纳点的生活垃圾、建筑垃圾的数量。统计时仅计算从生活垃圾、建筑垃圾源头和从生活垃圾转运站直接送到处理场和最终消纳点的清运量，对于二次中转的清运量不要重复计算。

餐厨垃圾属于生活垃圾的一部分，无论单独清运还是混合清运，都应统计在生活垃圾清运量中。

其中：**餐厨垃圾清运处置量**指单独清运，并且进行单独处置的餐厨垃圾总量，不含混在生活垃圾中清运的部分。

公共厕所

指供城市居民和流动人口使用，在道路两旁或公共场所等处设置的厕所。分为独立式、附属式和活动式三种类型。统计时只统计独立式和活动式，不统计附属式公厕。

独立式公共厕所按建筑类别应分为三类，活动式公共厕所按其结构特点和服务对象应分为组装厕所、单体厕所、汽车厕所、拖动厕所和无障碍厕所五种类别。

市容环卫专用车辆设备

指用于环境卫生作业、监察的专用车辆和设备，包括用于道路清扫、冲洗、洒水、除雪、垃圾粪便清运、市容监察以及与其配套使用的车辆和设备。如：垃圾车、扫路机（车）、洗路车、洒水车、真空吸粪车、除雪机、装载机、推土机、压实机、垃圾破碎机、垃圾筛选机、盐粉撒布机、吸泥渣车和专用船舶等。对于长期租赁的车辆及设备也统计在内。

统计时，单独统计道路清扫保洁专用车辆和生活垃圾运输专用车辆数。

村镇部分

人口密度

指建成区范围内的人口疏密程度。计算公式：

$$建成区人口密度（人 / 平方公里）＝ \frac{建成区常住人口（人）}{建成区面积（公顷）} \times 100$$

供水普及率

指报告期末建成区（村庄）用水人口与建成区（村庄）人口的比率。按建成区、村庄分别统计。计算公式：

$$建成区供水普及率（\%）＝ \frac{建成区用水人口（人）}{建成区常住人口（人）} \times 100\%$$

$$村庄供水普及率（\%）＝ \frac{村庄用水人口（人）}{村庄常住人口（人）} \times 100\%$$

燃气普及率

指报告期末建成区使用燃气的人口与建成区人口的比率。计算公式：

$$燃气普及率（\%）＝ \frac{建成区用气人口（人）}{建成区常住人口（人）} \times 100\%$$

人均道路面积

指报告期末建成区范围内平均每人拥有的道路面积。计算公式：

$$建成区人均道路面积（平方米 / 人）＝ \frac{建成区道路面积（万平方米）}{建成区常住人口（人）} \times 10^4$$

排水管道暗渠密度

指报告期末建成区范围内排水管道暗渠分布的疏密程度。计算公式：

$$建成区排水管道暗渠密度（公里/平方公里）=\frac{建成区排水管道长度（公里）+建成区排水暗渠长度（公里）}{建成区面积（公顷）}\times100$$

人均公园绿地面积

指报告期末建成区范围内平均每人拥有的公园绿地面积。计算公式：

$$建成区人均公园绿地面积（平方米/人）=\frac{建成区公园绿地面积（公顷）}{建成区常住人口（人）}\times10^{4}$$

建成区绿化覆盖率

指报告期末建成区范围内绿化覆盖面积与建成区面积的比率。计算公式：

$$建成区绿化覆盖率(\%)=\frac{建成区绿化覆盖面积（公顷）}{建成区面积（公顷）}\times100\%$$

建成区绿地率

指报告期末镇（乡）建成区范围内绿地面积与建成区面积的比率。计算公式：

$$建成区绿地率(\%)=\frac{建成区绿地面积（公顷）}{建成区面积（公顷）}\times100\%$$

人均日生活用水量

指用水人口平均每天的生活用水量。计算公式：

$$人均日生活用水量=报告期生活用水量/用水人口/报告期日历天数\times1000升$$

建成区

指行政区域内实际已成片开发建设、市政公用设施和公共设施基本具备的区域。建成区范围一般是指建成区外轮廓线所包括的地区，也就是实际建设用地达到的范围。建成区面积以镇人民政府建设部门（或规划部门）提供的范围为准。

村庄

指农村居民生活和生产的聚居点。

因为镇（乡）政府驻地和村庄的市政公用设施差距较大，所以在报表中，我们将镇、乡、农场等特殊区域分成建成区和村庄两部分进行统计。

村庄规划

指在镇总体规划或乡规划等指导下，根据经济发展水平，主要对住宅和供水、供电、道路、绿化、环境卫生以及生产配套设施建设的具体安排。村庄建设规划须经村民委员会议讨论同意，并由上级批准同意实施。统计时要注意规划的有效期，如至本年年底，规划已过期，则视为无规划。

本年村镇规划编制投入

指本年镇（乡）域内总体规划、专项规划、详细规划、村庄规划等各种规划编制工作的总投入。

公共供水

指公共供水企业以公共供水管道及其附属设施向单位和居民的生活、生产和其他各项建设提供用水。公共供水设施包括正规的水厂和虽达不到水厂标准,但不是临时供水设施,水质符合标准的其他公共供水设施。

综合生产能力

指按供水设施取水、净化、送水、输水干管等环节设计能力计算的综合生产能力。计算时，以四个环节中最薄弱的环节为主确定能力。没有设计能力的按实际测定的能力计算。对于经过更新改造后，实际生产能力与设计能力相差很大的，按实际能力填报。

供水管道长度

指从送水泵至用户水表之间直径 75mm 以上所有供水管道的长度。不包括从水源地到水厂的管道、水源井之间的联络管、水厂内部的管道、用户建筑物内的管道和新安装尚未使用的管道。按单管计算，即：如在同一条街道埋设两条或两条以上管道时，应按每条管道的长度相加计算。

年供水总量

指一年内供给建成区范围内的全部水量，包括有效供水量和损失的水量。既包括镇（乡）自建的供水设施供应的水量，也包括市（县）或其他乡镇供应的水量。

其中：**年生活用水量**指居民家庭与公共服务的年用水量。包括饮食店、医院、商店、学校、机关、部队等单位生活用水量，以及生产单位装有专用水表计量的生活用量（不能分开者，可不计）。

年生产用水量 指生产运营单位在生产、运营过程中的年用水量。

用水人口

指报告期末集中供水设施供给生活用水的家庭用户总人口数。可按本地居民户均人口数乘以家庭用水户数计算。

用气人口

指报告期末使用燃气（人工煤气、天然气、液化石油气）的家庭用户总人口数。可以本地居民平均每户人口数乘以燃气家庭用户数计算。

在统计液化气家庭用户数时，年平均用量低于 90 公斤的户数，忽略不计。

集中供热面积

指通过热网向建成区内各类房屋供热的房屋建筑面积。只统计供热面积达到 1 万平方米及以上的集中供热设施。

道路

指建成区范围内具有交通功能的各种道路。包括主干路、干路、支路、巷路。只统计路面宽度在 3.5 米（含 3.5 米）以上的各种铺装道路。

铺装道路路面包括水泥混凝土、沥青混凝土（柏油）、块石、砖块、混凝土预制块等形式。

道路长度、面积的计算

道路长度包括道路长度和与道路相通的桥梁、隧道的长度。道路面积只包括路面面积和与道路相通的广场、桥梁、停车场的面积，不含隔离带和绿化带面积。

污水处理厂

指污水通过排水管道集中于一个或几个处所，并利用由各种处理单元组成的污水处理系统进行净化处理，最终使处理后的污水和污泥达到规定要求后排放水体或再利用的生产场所。

污水处理厂主要有以下几种处理工艺：缺氧、好氧工艺（AO工艺）、2.厌氧、缺氧、好氧工艺（AAO工艺）、膜生物反应器工艺（MBR工艺）、氧化沟工艺（OD工艺）、序批式活性污泥工艺（SBR系列工艺）等。

氧化塘是污水处理的一种工艺，严格按《氧化塘设计规范》运行管理的氧化塘应作为污水处理厂。

污水处理装置

指在厂矿区设置的处理工业废水和周边地区生活污水的小型集中处理设备，以及居住区、度假村中设置的小型污水处理装置。

污水处理能力

指污水处理设施每昼夜处理污水量的设计能力,没有设计能力的按实际能力计算。

年污水处理总量

指一年内各种污水处理设施处理的污水量，包括将本地的污水收集到外地的污水处理设施处理的量，其中污水处理厂处理的量计为集中处理量。

绿化覆盖面积

指建成区内的乔木、灌木、草坪等所有植被的垂直投影面积。包括各种绿地的绿化种植覆盖面积、屋顶绿化覆盖面积以及零散树木的覆盖面积，不包括植物未覆盖的水域面积。

绿地面积

指报告期末建成区内用作园林和绿化的各种绿地面积。包括公园绿地、生产绿地、防护绿地、附属绿地的面积。其中：**公园绿地**指向公众开放的、以游憩为主要功能，有一定游憩设施的绿地。

生产绿地　指为绿化提供苗木、花草、种子的苗圃、花圃、草圃等圃地。

防护绿地　指用于安全、卫生、防风等功能的绿地。

附属绿地　指建设用地中绿地之外各类用地中的附属绿化用地。包括居住用地、公共设施用地、工业用地、仓储用地、对外交通用地、道路广场用地、市政设施用地和特殊用地中的绿地。

统计时注意单位是"公顷"，1 公顷=1 万平方米。

年生活垃圾处理量

指一年将建成区内的生活垃圾运到生活垃圾处理场（厂）进行处理的量，包括无害化处理量和简易处理量。

公共厕所数量

指在建成区范围内供居民和流动人口使用的厕所座数。统计时只统计独立的公厕，不包括公共建筑内附设的厕所。

房屋

按使用性质可划分为住宅、公共建筑和生产性建筑。

其中：**住宅**指坐落在村镇范围内，上有顶、周围有墙，能防风避雨，供人居住的房屋。按照各地生活习惯，可供居住的帐篷、毡房、船屋等也包括在内，兼作生产用房的房屋可以算为住宅。包括厂矿、企业、医院、机关、学校的集体宿舍和家属宿舍，但不包括托儿所、病房、疗养院、旅馆等具有专门用途的房屋。

公共建筑　指坐落在村镇范围内的各类机关办公、文化、教育、医疗卫生、商业等公共服务用房。

生产性建筑　指坐落在村镇范围内的包括乡镇企业厂房、养殖厂、畜牧场等用房及各类仓库用房的建筑面积。

本年竣工建筑面积

指本年内完工的住宅（公共建筑、生产性建筑）建筑面积。包括新建、改建和扩建的建筑面积，未完工的在建房屋不要统计在内。统计时按镇（乡）建成区和村庄分开统计。

其中：**房地产开发**建筑面积，指房地产开发企业，按照城乡建设规划要求，立项审批（备案）并取得《施工许可证》后，在依法取得土地使用权的土地上开发的楼盘或小区工程项目的建筑面积。

年末实有建筑面积

指本年年末，坐落在村镇范围内的全部住宅（公共建筑、生产性建筑）建筑面积。统计时按镇（乡）建成区和村庄分开统计。计算方法为：

年末实有房屋建筑面积=上年末实有+本年竣工+本年区划调整增加-本年区划调整减少

-本年拆除（倒塌、烧毁等）

混合结构以上

指结构形式为混合结构及其以上（如钢筋混凝土结构、砖混结构）的房屋建筑面积。

Explanatory Notes on Main Indicators

Indicators in the statistics for Cities and County Seats

Population Density

It refers to the density of population in a given zone. The calculation equation is:

$$\text{Population Density} = \frac{\text{Population in Urban Areas} + \text{Urban Temporary Population}}{\text{Urban Area}}$$

Daily Domestic Water Use Per Capita

It refers to average amount of daily water consumed by each person. The calculation equation is:

Daily Domestic Water Use Per Capita＝（Water Consumption by Households + Water Use for Public Service+Domestic water consumption in the free water supply）÷Population with Access to Water Supply ÷Calendar Days in Reported Period× 1000 liters

Water Coverage Rate

It refers to proportion of urban population supplied with water to urban population. The calculation equation is:

$$\text{Water Coverage Rate} = \frac{\text{Urban Population with Access to Water Supply (Including Temporary Population)}}{\text{Urban Permanent Population} + \text{Urban Temporary Population}} \times 100\%$$

$$\text{Public Water Coverage Rate} = \frac{\text{Urban Population with Access to Public Water Supply (Including Temporary Population)}}{\text{Urban Permanent Population} + \text{Urban Temporary Population}} \times 100\%$$

Gas Coverage Rate

It refers to the proportion of urban population supplied with gas to urban population. The calculation equation is:

$$\text{Gas Coverage Rate} = \frac{\text{Urban Population with Access to Gas (Including Temporary Population)}}{\text{Urban Permanent Population} + \text{Urban Temporary Population}} \times 100\%$$

Surface Area of Urban Roads Per Capita

It refers to the average surface area of urban roads owned by each urban resident at the end of reported period. The calculation equation is:

$$\text{Surface Area of Roads Per Capita} = \frac{\text{Surface Area of Roads in Given Urban Areas}}{\text{Urban Permanent Population} + \text{Urban Temporary Population}}$$

Density of Road Network in Built Districts

It refers to the extent which roads cover built districts at the end of reported period. The calculation equation is:

$$\text{Density of Road Network in Built Districts} = \frac{\text{Length of Roads in Built Districts}}{\text{Floor Area of Built Districts}}$$

Density of Drainage Pipelines in Built Districts

It refers to the extent which drainage pipelines cover built districts at the end of reported period. The calculation equation is:

$$\text{Density of Drainage Pipelines in Built Districts} = \frac{\text{Length of Drainage Pipelines in Built Districts}}{\text{Floor Area of Buil Districts}}$$

Wastewater Treatment Rate

It refers to the proportion of the quantity of wastewater treated to the total quantity of wastewater discharged at the end of reported period. The calculation equation is:

$$\text{Wastewater Treatment Rate} = \frac{\text{Quantity of Wastewater Treated}}{\text{Quantity of Wastewater Discharged}} \times 100\%$$

Centralized Treatment Rate of Wastewater Treatment Plants

It refers to the proportion of the quantity of wastewater treated in wastewater treatment plants to the total quantity of wastewater discharged at the end of reported period. The calculation equation is:

$$\text{Centralized Treatment Rate of Wastewater Treatment Plants} = \frac{\text{Quantity of wastewater treated in wastewater treatment facility}}{\text{Quantity of wastewater discharged}} \times 100\%$$

Public Recreational Green Space Per Capita

It refers to the average public recreational green space owned by each urban dweller in given areas. The calculation equation is:

$$\text{Public Recreational Green Space Per Capita} = \frac{\text{Public green space in given urban areas}}{\text{Urban Permanent Population} + \text{Urban Temporary Population}}$$

Green Coverage Rate of Built Districts

It refers to the ratio of green coverage area of built districts to surface area of built districts at the end of reported period. The calculation equation is:

$$\text{Green Coverage Rate of Built Districts} = \frac{\text{Green Coverage Area of Built Districts}}{\text{Area of Built Districts}} \times 100\%$$

Green Space Rate of Built Districts

It refers to the ratio of area of parks and green land of built districts to the area of built up districts at the end of reported period. The calculation equation is:

$$\text{Green Space Rate of Built Districts} = \frac{\text{Area of parks and green land of built districts}}{\text{Area of Built Districts}} \times 100\%$$

Domestic Garbage Treatment Rate

It refers to the ratio of quantity of domestic garbage treated to quantity of domestic garbage produced at the end of reported period. The calculation equation is:

$$\text{Domestic Garbage Treatment Rate} = \frac{\text{Quantity of Domestic Garbage Treated}}{\text{Quantity of Domestic Garbage Produced}} \times 100\%$$

Domestic Garbage Harmless Treatment Rate

It refers to the ratio of quantity of domestic garbage treated harmlessly to quantity of domestic garbage produced at the end of reported period. The calculation equation is:

$$\text{Domestic Garbage Harmless Treatment Rate} = \frac{\text{Quantity of Domestic Garbage Treated Harmlessly}}{\text{Quantity of Domestic Garbage Produced}} \times 100\%$$

Urban （county） District Area

It refers to the total land area （including water area） under jurisdiction of cities（counties）.

Urban （county） Area

It refers to the area of a city's urban and county area.

Urban area of a city includes: （1） City-level areas administered by neighborhood office; （2） other towns （villages） connected to city public facilities, residential facilities and municipal utilities; （3） Independent Industrial and Mining District, Development Zones, special areas like research institutes , universities and colleges with permanent residents of 3000 above.

County area of a city includes: （1） Town or areas administered by neighborhood office; （2） other towns （villages） connected to county public facilities, residential facilities and municipal utilities; （3） Independent Industrial and Mining District, Development Zones, special areas like research institutes , universities and colleges with permanent residents of 3000 above in county.

Towns （villages） connected to city public facilities, residential facilities and municipal utilities means that towns and urban centers are connected with public facilities, residential facilities and municipal utilities that are constructed or under construction, and not cut off by non-construction land like water area, agricultural land, parks, woodland or pasture.

As to conurbation or cities organized in scattered form, urban area should include the separated or decentralized area.

Urban （county） Permanent Population

It refers to permanent residents in urban （county） areas, which are in compliance with the number of permanent residents registered in public security authorities.

Temporary Population

Temporary population includes people who leave permanent address, and live for over half a year in a place. Counted by urban area, county, urban area and county seat,which are in compliance with the number of temporary residents registered in public security authorities.

Area of Built District

It refers to large scale developed quarters within city （county） jurisdiction with basic public facilities and utilities. For a nucleus city, the built-up district consists of large-scale developed quarters with basic public facilities and utilities, which are either centralized or decentralized. For a city with several towns, the built-up district consists of several developed quarters attached in succession with basic public facilities and utilities. Range of built-up district is the area encircled by the limits of the built-up district, i.e. the range within which the actual developed land of this city exists.

Area of Urban Land for development Purpose

It refers to area of land including land for Residential Development, Administration and Public Services, Commercial and Business Facilities, Industrial, Manufacturing, Logistics and Warehouse, Road, Street and Transportation, Municipal Utilities, Green Space and Square.Separately counted by the planned construction land and the current construction land.

The area of land used for planned construction refers to the area of land used for cities （towns） as determined by the latest version of the overall urban plans approved according to law for the towns where the relevant cities and county people's governments are located at the end of the reporting period.

The area of current land for construction purposes refers to the area corresponding to the actual situation of urban land for construction purposes in the towns where the relevant cities and county people's governments are located at the end of the reporting period.

Urban Maintenance and Construction Fund

It refers to fund used for urban construction and maintenance, including urban maintenance and construction tax, extra-charges for public utilities, financial allocation from the central government and local governments, domestic loan, securities revenue, foreign investment, revenue from land transfer and assets replacement, self-raised fund by municipal utilities, charges by administrative and institutional units for urban maintenance and construction, pooled revenue and other revenues.

Urban Maintenance and Construction Tax

It is one of local taxes imposed according to *Temporary Regulations of People's Republic of China on Urban Maintenance and Construction Tax*. It is levied based on actual value-added tax, consumption tax and business tax paid by a taxpayer. It is also paid with value-added tax, consumption tax and business tax at the same time. Different rates apply to different places. The rate comes to 7% in urban areas, 5% in counties and towns and 1% beyond these places.

Extra Charge from Urban Utilities

It is additional revenue obtained from additional fees imposed on provision of products and service of such urban utilities as power generation, water supply for the industry, operation of public bus and trolley bus, taped water supply, electricity for lightening, telephone operation, gas and ferry etc. for the purpose of urban construction and maintenance.

Investment in Fixed Assets

It is the economic activities featuring construction and purchase of fixed assets, i.e. it is an essential means for social reproduction of fixed assets. The process of reproducing fixed assets includes fixed assets renovation （part and full renovation）, reconstruction, extension and new construction etc. According to the new industrial financial accounting system, cost of major repairs for part renovation of fixed assets is covered by direct cost. According to the current investment management and administrative regulations, any repair and maintenance works are not included in statistics as investment in fixed assets. Innovation projects on current municipal service facilities should be included in statistics.

Newly Added Fixed Assets

They refer to the newly increased value of fixed assets, including investment in projects completed and put into operation in the reported period, and investment in equipment, tools, vessels considered as fixed assets as well as relevant expenses should be included in. Other construction expenses that increase the value of fixed assets should be included in the newly added fixed assets along with the project delivered for use.

Newly Added Production Capacity （or Benefits）

It refers to newly added design capacity through investment in fixed assets. Newly added production capacity or benefits is calculated based on projects which can independently produce or bring benefits once projects are put into operation.

Integrated Production Capacity

It refers to a comprehensive capacity based on the designed capacity of components of the process, including water collection, purification, delivery and transmission through mains. In calculation, the capacity of weakest component is the principal determining capacity. For the updated and reformed, filling in the new design capability after the updated and reformed.

Length of Water Pipelines

It refers to the total length of all pipes from the pumping station to individual water meters. If two or more pipes line in parallel in a same street, the length of water pipelines is the length sum of each line.

Total Quantity of Water Supplied

It refers to the total quantity of water delivered by water suppliers during the reported period, including accounted water and unaccounted water.

Accounted water refers to the actual quantity of water delivered to and used by end users, including water sold and free.

Unaccounted Water

It refers to sum of water leakage due to malfunction of water meters, measurement of water loss and other loss of water. The measurement of loss includes the total water loss for residential users and the water loss with table error for non-residential users; Other loss of water refers to the loss of water caused by management factors such as unregistered users' water and users' refusal to check.

Quantity of Fresh Water Used

It refers to the quantity of water obtained from any water source for the first time, including tap water, groundwater, surface water. As for a city, it includes the quantity of fresh water used by urban water suppliers and customers in different industries and sectors.

Among which, **the amount of new industrial water taken** refers to the amount of fresh water that is actually drawn from various water sources and used for any purpose in order to ensure the normal progress of industrial production and the need for water in the production process, including the amount of indirect cooling water and process water, new boiler water volume and other new water volume.

Quantity of Recycled Water

It refers to the sum of water recycled, reclaimed and reused by customers.

Among which, **the amount of recycled industrial water** refers to the sum of the amount of recycled water used in the domestic and production water of industrial enterprises and the amount of water recovered and reused directly or after treatment.

Quantity of Water Saved

It refers to the water saved in the reported period through efficient water saving measures, e.g. improvement of production methods, technologies, equipment, water use behavior or replacement of defective and inefficient devices, or strengthening of management etc.

Integrated Gas Production Capacity

It refers to the combined capacity of components of the process such as gas production, purification and delivery with an exception of the capacity of backup facilities at the end of reported period. It is usually calculated based on the design capacity. Where the actual capacity surpluses the design capacity, integrated capacity should be calculated based on the actual capacity, mainly depending on the capacity of the weakest component.

Length of Gas Supply Pipelines

It refers to the length of pipes operated in the distance from a compressor's outlet or a gas station exit to pipes connected to individual households at the end of reported period, excluding pipes through coal gas production plant, delivery station, LPG storage station, bottled station, storage and distribution station, air mixture station, supply station and user's building.

Quantity of Gas Supplied

It refers to amount of gas supplied to end users by gas suppliers at the end of reported period. It includes sales amount and loss amount.

Gas Stations for Gas-Fueled Motor Vehicles

They are designated stations that provide fuels such as compressed natural gas, LPG to gas-fueled motor vehicles. Statistics should be based on different gas types.

Heating Capacity

It refers to the designed capacity of heat delivery by heat suppliers to urban customers.

Total Quantity of Heat Supplied

It refers to the total quantity of heat obtained from steam and hot water, which is delivered to urban users by heat suppliers during the reported period.

Length of Heating Pipelines

It refers to the total length of pipes for delivery of steam and hot water from heat sources to entries of buildings, excluding lines within heat sources.

Among which, **the primary pipe network** refers to the heating pipeline from the heat source to the heating station, **the secondary pipe network** refers to the heating pipeline from the heating station to the user.

Urban roads

They refer to roads, bridges, tunnels and auxiliary facilities that are provided to vehicles and passengers for transportation. Urban roads consist of drive lanes and sidewalks. Only paved roads with width with and above 3.5 m, including roads within on-limits industrial parks and residential communities are included in statistics.

Length of Roads

It refers to the length of roads and bridges and tunnels connected to roads. It is calculated based on the length of centerlines of traffic lanes.

Surface Area of Roads

It refers to surface area of roads and squares, bridges and tunnels connected to roads（surface area of roadway and sidewalks are calculated separately）. Surface area of sidewalks is the area sum of each side of sidewalks including pedestrian streets and squares, excluding car-and-pedestrian mixed roads.

Bridges

They refer to constructed works that span natural or man-made barriers. They include bridges spanning over rivers, flyovers, overpasses and underpasses.

Number of Road Lamps

It refers to the sum of road lamps only for illumination, excluding road lamps for landscape lightening.

Length of Flood Control Dikes

It refers to actual length of constructed dikes, which sums up the length of dike on each bank of river. Where each bank has more than one dike, only the longest dike is calculated.

Quantity of Wastewater Discharged

It refers to the total quantity of domestic sewage and industrial wastewater discharged, including effluents of sewers, ditches, and canals.

(1) It is calculated by multiplying the daily average of measured discharges from sewers, ditches and canals with calendar days during the reported period.

(2) If there is a device to read actual quantity of discharge, it is calculated based on reading.

(3) If without such ad device, it is calculated by multiplying total quantity of water supply with wastewater drainage coefficient.

Category of urban wastewater	Wastewater drainage coefficient
Urban wastewater	0.7-0.8
Urban domestic wastewater	0.8-0.9
Urban industrial wastewater	0.7-0.9

Length of Drainage Pipelines

The total length of drainage pipelines is the length of mains, trunks, branches plus distance between inlet and outlet in manholes and junction wells. The calculation should be based on a single pipe, that is, if there are two or more drainage pipes side by side on the same street, the calculation should be based on the length of each drainage pipe.

Among which, **Sewage Pipe** refers to a drainage pipe dedicated to discharge sewage.

Rainwater Pipe refers to drainage pipes dedicated to draining rainwater.

Rain and Sewage Confluence Pipeline refers to the drainage pipeline where rainwater and sewage enter the same pipeline for drainage at the same time.

Quantity of Wastewater Treated

It refers to the actual quantity of wastewater treated by wastewater treatment plants and other treatment facilities in a physical, or chemical or biological way.

Among which, **the treatment outside the city （county）** refers to the sewage treatment plant as a regional facility that not only treats the sewage of the city （county）, but also treats the sewage of other cities, counties, or other villages outside the city （county）. This part of the sewage treatment volume is separately counted and deducted when calculating the sewage treatment rate of the city （county）.

Dry Sludge Production

It refers to the final quantity of dry sludge produced in the process of wastewater treatment plants. Dry sludge refers to the amount of sludge calculated by the mass of dry solids, with a moisture content of 0. If the moisture content of the produced wet sludge is n%, then the amount of dry sludge produced = the amount of wet sludge produced $\times$ （1-n%）.

Dry Sludge Disposal Volume

It refers to the amount of dry sludge to be treated and disposed according to the standard during the reporting period. Fill in the statistics according to land use, building materials use, incineration, landfill and others.

Among which, **Sludge Land Utilization** refers to the disposal of sludge products after the treatment of the standard for landscaping, land improvement, woodland, agriculture and other occasions.

The Utilization of Sludge Building Materials refers to the disposal of the products after sludge treatment reaches the standard as part of the raw materials of building materials such as bricks and cement clinker.

Sludge Incineration refers to the treatment and disposal methods that use incinerators to completely mineralize sludge into a small amount of ashes, including individual incineration, and co-incineration with industrial kilns such as domestic waste and thermal power plants.

Land Filling of Sludge refers to a disposal method in which engineering measures are taken to pile, fill, and bury the sludge products that have reached the treatment standards, and place them in a controlled site.

Green Coverage Area

It refers to the vertical shadow area of vegetation such as trees （arbors）, shrubs and grasslands. It is the the vertical shadow area sum of green space, roof greening and scattered trees coverage. The vertical shadow area of bushes and herbs under the shadow of trees and herbs under the shadow of shrubs should not be counted repetitively.

Area of Green Space

It refers to the area of all spaces for parks and greening at the end of reported period, which includes area of public recreational green space, shelter belt, land for squares, attached green spaces, and the area of regional green space in built district.

Among which, **Public Recreational Green Space** refers to green space with recreation and service facilities. It also serves some comprehensive functions such as improving ecology and landscape and preventing and mitigating disasters.

Shelter Belt refers to the green space that is independent of land and has the functions of hygiene, isolation, safety, and ecological protection, and is not suitable for tourists to enter. It mainly includes protective green space for health isolation, protective green space for roads and railways, protective green space for high-pressure corridors and protective green space for public facilities.

Land for Squares refer to the urban public activity space with the functions of recreation, memorial, gathering and avoiding danger.

Attached Green Spaces refer to green land attached to various types of urban construction land （except "green land and square land"）. Including residential land, land for public management and public service facilities, land for commercial service facilities, industrial land, land for logistics and storage, land for roads and transportation facilities, land for public facilities and other green spaces.

Regional Green Space refers to the green land located outside the urban construction land, which has the functions of urban and rural ecological environment, natural resources and cultural resources protection, recreation and fitness, safety protection and isolation, species protection, garden seedling production and so on.

Parks

They refer to places open to the public for the purposes of tourism, appreciation, relaxation, and undertaking scientific, cultural and recreational activities. They are fully equipped and beautifully landscaped green spaces. There are different kinds of parks, including general park, Children Park, park featuring historic sites and culture relic, memorial park, scenic park, zoo, botanic garden, and belt park. Recreational space within communities is not included in statistics, while only general parks, theme parks and belt parks at city or district level are included in statistics.

Among which, **Free-ticket Parks** refer to parks that are free to the public and do not sell tickets.

Area of Roads Cleaned and Maintained

It refers to the area of urban roads and public places cleaned and maintained at the end of reported period, including drive lanes, sidewalks, drive tunnels, underpasses, green space attached to roads, metro stations, elevated roads, flyovers, overpasses, squares, parking areas and other facilities. Where a place is cleaned and maintained several times a day, only the time with maximum area cleaned is considered.

Among which, **Mechanized Road Cleaning and Cleaning Area** refers to the road area that was cleaned and cleaned with large and small machinery such as road sweepers （machines） and washing vehicles at the end of the reporting period. When multiple machines are repeatedly used on a road, only the area cleaned and cleaned by one machine is calculated, and the statistics cannot be repeated.

Quantity of Domestic Garbage and Construction Waste Transferred

It refers to the total quantity of domestic garbage and construction waste collected and transferred to treatment grounds. Only quantity collected and transferred from domestic garbage and construction waste sources or from domestic garbage transfer stations to treatment grounds is calculated with exception of quantity of domestic garbage and construction waste transferred second time.

Food waste is a part of domestic waste. Whether it is removed separately or mixedly, it should be counted in the volume of domestic waste.

Among which, **the amount of food waste removal and disposal** refers to the total amount of food waste that is separately removed and disposed of separately, excluding the part that is mixed in the domestic waste.

Latrines

They are used by urban residents and flowing population, including latrines placed at both sides of a road and public place. They usually consist of detached latrine movable latrine and attached latrine. Only detached latrines rather than latrines attached to public buildings are included in the statistics.

Detached latrine is classified into three types. Moveable latrine is classified into fabricated latrine, separated latrine, motor latrine, trail latrine and barrier-free latrine.

Specific Vehicles and Equipment for Urban Environment Sanitation

They refer to vehicles and equipment specifically for environment sanitation operation and supervision, including vehicles and equipment used to clean, flush and water the roads, remove snow, clean and transfer garbage and soil, environment sanitation monitoring, as well as other vehicles and equipment supplemented, such as garbage truck, road clearing truck, road washing truck, sprinkling car, vacuum soil absorbing car, snow remover, loading machine, bulldozer, compactor, garbage crusher, garbage screening machine, salt powder sprinkling machine, sludge absorbing car and special shipping. Vehicles and equipment for long-term lease are also included in the statistics.

The number of special vehicles for road cleaning and cleaning and the special vehicles for domestic garbage transportation are separately counted.

Indicators in the Statistics for Villages and Small Towns

Population density

Population density is the measure of the population per unit area in Built districts.

Calculation equation is:

Population Density in Built Districts（People per square kilometer）= Permanent Population in Built Districts（People）×100/ Area of Built Districts（Hectares）

Water Coverage Rate

It refers to proportion of population supplied with water in built districts （villages） to population in built districts （villages） at the end of reported period. Statistics is made by built district and village respectively. The calculation equation is:

Water Coverage Rate in Built Districts（%） = Population with Access to Water Supply in Built Districts （People）/ Permanent Population in Built Districts（People）×100%

Water Coverage Rate in Villages（%） = Population with Access to Water Supply in Villages（People）/ Permanent Population in Village（People）×100%

Gas Coverage Rate

It refers to the proportion of population supplied with gas in built districts to urban population in built districts at the end of reported period. The calculation equation is:

Gas Coverage Rate（%）=Population Supplied with Gas in Built Districts（People）/Permanent Population in Built Districts（People）×100%

Surface Area of Roads Per Capita

It refers to the average surface area of roads owned by each urban resident in built district at the end of reported period. The calculation equation is:

Surface Area of Roads in Built Districts Per Capita（Square meters/person）= Surface Area of Roads in Built Districts（Ten thousand square meters）×10^4 / Permanent Population in Built Districts（people）

Density of Drainage Pipeline Culverts

It refers to the extent which drainpipes cover built districts at the end of reported period. The calculation equation is:

Density of Drainage Pipeline Culverts in Built Districts（Kilometers/square kilometers）=Length of Drainage Pipelines in Built Districts（Kilometers）+Length of Drainage Culverts in Built Districts（Kilometers）×100/Floor Area of Built Districts（Hectares）

Public Recreational Green Space Per Capita

It refers to the average public recreational green space owned by each dwell in built district at the end of reported period. The calculation equation is:

Public Recreational Green Space in Built Districts Per Capita（Square meters/person）= Area of Public Recreational Green Space in Built Districts（Hectares）×10^4/ Permanent Population in Built Districts（people）

Green Coverage Rate in Built Districts

It refers to the ratio of green coverage area of built districts to surface area of built district at the end of reported period. The calculation equation is:

Green Coverage Rate in Built Districts（%）= Built Districts Green Coverage Area（Hectares）/ Area of Built Districts（Hectares）×100%

Green Space Rate in Built Districts

It refers to the ratio of area of parks and green land of built districts to the area of built districts at the end of reported period. The calculation equation is:

Green Space Rate in Built Districts（%）=Built Districts Area of Parks and Green Space（Hectares）/Area of Built Districts（Hectares）×100%

Daily Domestic Water Use Per Capita

It refers to average amount of daily water consumed by each person. The calculation equation is:

Domestic Water Use Per Capita= Domestic Water Consumption / Population with Access to Water Supply / Calendar Days in Reported Period× 1000 liters

Built District

It refers to large scale developed quarters within city jurisdiction with basic public facilities and utilities, The scope of the built-up area generally refers to the area covered by the outline of the built-up area, that is, the range reached by the actual construction land which is subject to area verified by local construction authorities or planning authorities.

Village

It refers to a collection of houses and other buildings in a country area where rural residents live and produce.

Because there is a big gap between the town （township） government station and the municipal public facilities in the village, we divide the special areas such as town, township and farm into built-up areas and villages for statistics in the report form.

Villages Planning

It addresses, under guidance of town overall planning or township planning, detailed arrangement for residential buildings, water supply, power supply, roads, greening, environmental sanitation and supplementary facilities for production according to local economic development level. Development Planning in Villages must be approved by the superior agreed to implement. Special attention should be given to planning validation in statistics.

Input in Village and Town Planning Development this Year

It refers to the total investment in the preparation of various plans such as overall planning, special planning, detailed planning, and village planning in the town （township） this year.

Public Water Supply

It refers to water supply for production, domestic use as well as public service by public water suppliers through water supply pipes and accessories. Public water supply facilities include regular plants , and other facilities which are not temporary facilities and produce water in compliance with standards.

Integrated Water Production Capacity

It refers to a comprehensive capacity based on the design capacity of components of the process, including water collection, purification, delivery and transmission through mains. In calculation, the capacity of weakest component is the principal determining capacity. Without design capacity, it is calculated based on capacity measured. If real production capacity differs widely from design capacity through renovation, real production capacity will be the basis for calculation.

Length of Water Pipelines

It refers to the total length of all pipes from the pumping station to individual water meters, excluding pipes from water resources to water plants, connection pipes between water resources wells, pipes within water plants and end-use buildings as well as pipes newly installed but not uses yet. If two or more pipes line in parallel in a same street, the length of water pipelines is the length sum of each line.

Total Annual Water Supply

It refers to the total amount of water supplied to the built-up area within a year, including the effective water supply and the amount of water lost. It includes not only the amount of water supplied by the self-built water supply facilities of the town （township）, but also the amount of water supplied by the city （county） or other towns.

Among which, **Yearly Domestic Water Use** refers to amount of water consumed by households and public facilities, including domestic water use by restaurant, hospital, department store, school, government agency and army as well as water metered by special meters installed in industrial users for domestic use.

Yearly Water Consumption for Production refers to water consumed each year by industrial users for production and operation.

Population with Access to Water

It refers to the total number of household users whose water was supplied by centralized water supply facilities at the end of the reporting period. It can be calculated by multiplying the number of household water users by the average number of family members per household.

Population with Access to Gas

It refers to total quantity of customers in households supplied with gas （including man-made gas, natural gas and LPG） at the end of reported period.

When comes to number of households supplied with LPG, the households with yearly average use of LPG being below 90 kilograms are not included in statistics.

Area of Centrally Heated District

It refers to floor area of building stock in built-up districts supplied with centralized heating system. Only areas with centrally heated floor space of 10,000 square meters and above are included in statistics.

Road

It refers to a variety of roads with traffic functions within the built-up area. Including trunk road, trunk road, branch road, lane road. Roads whose width is 3.5 meters or over are included in statistics.

Pavement road surface includes cement concrete, asphalt concrete （asphalt）, block stone, brick, concrete precast blocks and other forms.

Length and Surface Area of Road

It refers to the length of roads and bridges and tunnels connected to roads. It is calculated based on the length of centerlines of traffic lanes. Surface area of roads only includes surface area of roads and squares, bridges and tunnels connected to roads, excluding separation belt and green belt.

Wastewater Treatment Plant

It refers to places where wastewater is collected in one or a few places through drainage pipes, then purified and treated by wastewater treatment system with treated water and sludge being discharged or recycled in accordance with related standards and requirements.

Sewage treatment plants mainly have the following treatment processes: anoxic, aerobic process （AO process）, 2. Anaerobic, anoxic, aerobic process （AAO process）, membrane bioreactor process （MBR process）, oxidation ditch process （OD process）, sequence batch activated sludge process （SBR series process）, etc.

Oxidation pond is one of techniques applied in wastewater treatment. Oxidation pond which is strictly operated in accordance with Oxidation Pond Design Standard should be employed as wastewater treatment plant

Wastewater Treatment Facilities

They refer to small-size and centralized equipment and devices installed in plant or mining area to treat industrial wastewater and domestic wastewater in surrounding areas , as well as wastewater equipment and devices placed in residential areas.

Wastewater Treatment Capacity

It refers to designed capacity for amount of wastewater treated every day and night by wastewater treatment facilities. If there is lack of design capacity, real capacity will be the determining factor for calculation.

Yearly Amount of Wastewater Treated

It refers to amount of wastewater treated by different wastewater treatment facilities within a year, including the amount of local wastewater collected and delivered to wastewater treatment facilities in other areas. The amount of wastewater treated in wastewater treatment plants are regarded as centralized treated amount.

Green Coverage Area

It refers to the vertical shadow area of vegetation such as trees （arbors）, shrubs and grasslands. It is the area sum of green space, roof greening and scattered trees coverage. The surface area of water bodies which are not covered by green is not included.

Area of Green Space

It refers to the area of all spaces for parks and greening at the end of reported period, which includes area of public recreational green space, green land for production, shelter belt, and attached green spaces.

Among which, **Public Recreational Green Space** refers green space equipped with recreational facilities and open to the public for recreational purposes.

Green Land for Production refers to the nursery, flower and grass beds that provide seedlings, flowers and seeds for greening.

Shelter Belt refers to the green space used for safety, health, wind protection and other functions.

Attached Green Spaces refer to the ancillary green land in all kinds of land other than green land in the construction land. Including residential land, land for public facilities, land for industry, land for storage, land for external traffic, land for roads and squares, land for municipal facilities and green land in special land.

Note that the unit of statistics is "hectare",1 hectare = 10,000 square meters.

Yearly Amount of Domestic Garbage Treated

It refers to amount of domestic garbage in built district which are delivered to domestic garbage treatment plants for harmless treatment or simplified treatment each year.

Number of Latrines

It refers to number of latrines for residents and floating population. Statistics only includes detached latrines, and latrines located within buildings are not included.

House

According to the nature of use, it can be divided into residential building, public building and building for production.

Among which, **Residential Building** refers to residential building with roof and wall that shelters people from external environment. In light of the local customs, houses also include tent, yurt and ship where people could live, building used for production in addition to habitation, and living quarters for workers and staff supplied by factories and mines, enterprises, hospitals, government agencies and schools. Nursery, ward, sanatorium, and hotel which are special purpose buildings are not included.

Public Building refers to buildings for government agencies, culture activities, healthcare, and business and commerce, which are located in towns and villages.

Building for Production refers to buildings for township enterprises, farms, stock farms, and warehouses located in towns and villages.

Floor Area of Buildings Completed this Year

It refers to floor area of buildings （public buildings and buildings for production） completed in this year, including floor area of new construction, renovation, and extension. Ongoing construction is not included in statistics. Data are separately collected for the built-up districts of small towns and villages.

Among which, **Real Estate Development Building Area** refers to the construction of real estate or community engineering projects developed on land where land use rights are legally obtained by real estate development enterprises, in accordance with the requirements of urban and rural construction planning, after project approval （recording） and obtaining the Construction Permit area.

Real Floor Area of Buildings at the End of Year

It refers to floor area of total buildings （public buildings and buildings for production） located in towns and villages at the end of year. Calculation equation:

Real Floor Area of Buildings at the End of Year = Real Floor Area of Buildings at The End of Last Year+ Added Floor Area of Buildings Due to District Readjustment This Year- Reduced Floor Area of Buildings Due to District Readjustment This Year-floor Area of Buildings Dismantled （collapsed or burned down） This Year.

Building with Mixed Structure

It refers to floor area of buildings with mixed structure, such as reinforced concrete structure, brick-and-concrete structure etc.